Phase Equilibria in Ionic Liquid Facilitated Liquid–Liquid Extractions

Phase Equilibria in Ionic Liquid Facilitated Liquid–Liquid Extractions

Anand Bharti, Debashis Kundu,
Dharamashi Rabari and Tamal Banerjee

CRC Press is an imprint of the
Taylor & Francis Group, an **informa** business
AN AUERBACH BOOK

CRC Press
Taylor & Francis Group
6000 Broken Sound Parkway NW, Suite 300
Boca Raton, FL 33487-2742

First issued in paperback 2020

ISBN 13: 978-0-367-57379-9 (pbk)
ISBN 13: 978-1-4987-6948-8 (hbk)

Library of Congress Cataloging-in-Publication Data

Names: Bharti, Anand.
Title: Phase equilibria in ionic liquid facilitated liquid-liquid extractions / Anand Bharti [and three others].
Description: Boca Raton, FL : CRC Press, Taylor & Francis Group, [2017] | Includes bibliographical references and index.
Identifiers: LCCN 2016041214| ISBN 9781498769488 (hardback) | ISBN 9781315367163 (ebook)
Subjects: LCSH: Extraction (Chemistry) | Liquid-liquid interfaces. | Liquid-liquid equilibrium. | Ionic solutions.
Classification: LCC TP156.E8 P43 2017 | DDC 660/.2842--dc23
LC record available at https://lccn.loc.gov/2016041214

Visit the Taylor & Francis Web site at
http://www.taylorandfrancis.com

and the CRC Press Web site at
http://www.crcpress.com

Contents

List of Figures

List of Tables

Preface

Background

Ionic liquids (ILs) are a new class of nonvolatile solvents with a nonsymmetric organic cation and an organic or inorganic anion. For the past couple of decades, ILs had been considered as a better replacement for conventional volatile solvents. The rate of publication in the open literature has been rising at a higher rate for years. A total of 12,400 papers have been published since 2007 against 7,400 published between 1990 and March 2007. These data show the increase in new applications of ILs in various conventional and nonconventional domains such as catalysis, nanotechnology, azeotropic separation, liquid–liquid extraction, metal extraction, membrane science and gas absorption. This book contains a detailed discussion of the calculation methods in the multi-scale modelling of mixtures containing ILs. It relates the physical properties of ILs with the phase behavior of an IL-containing system. Finally it also compares the results with the nontraditional optimization technique so as to scale up the laboratory extraction process to an industrial scale.

Objectives

This book is intended for the use of graduate students, industry professionals, researchers and academicians of a diverse community of chemical engineers, process engineers, material scientists, biochemical engineers and chemists. It may be used as a reference book for practical problems that arise in phase equilibria calculations and process optimization. The objective of this book is to present an extensive discussion on the following topics:

- Experimental techniques and model prediction of complex phase behavior involving IL systems. State-of-the-art liquid–liquid equilibria and vapor–liquid–liquid equilibria problems will give readers a wealth of real-world examples using thermodynamic problems.
- Convergence of quantum chemical prediction and statistical mechanical methods will help us develop an intuitive understanding of separation processes from the molecular to the bulk level.

- State-of-the-art optimization techniques will give readers insights to scale up an implementation from the laboratory to the industrial level.

We envisage that with clear explanations of the concepts, the readers will acquire the necessary skills for phase equilibria calculations.

Motivation and Goal

An overwhelming number of publications in phase equilibria and IL-containing systems motivate us to propose this book where experiments and characterization techniques along with model-based predictions will give a ready-to-use database for engineers and researchers. Our goals of writing this book are as follows:

- Communicating directly to engineers and researchers in a simple yet organized manner.
- Providing pathways for leading engineers and researchers towards a clear understanding and firm grasp of the phase equilibria calculations.
- Encouraging creative thinking and applications of predictive models for other types of engineering problems.

Overview of Chapters

The topics of the chapters are summarized as follows:

- Chapter 1 is a general introduction of ILs and the phase behavior of IL-containing systems. The nature of the systems considered is briefly introduced along with optimization techniques such as Genetic Algorithm (GA), particle swarm optimization (PSO) and Cuckoo Search (CS). It also discusses predictive models such as conductor-like screening model segment activity coefficients (COSMO-SAC) and perturbed chain-statistical associating fluid theory (PC-SAFT) equation of state.
- Chapter 2 reports LLE experiments along with correlation of LLE data using excess Gibbs free energy models. It also discusses application of

predictive models such as PC-SAFT to LLE. Representative examples such as extraction of biofuels and biochemicals has been discussed.
- Chapter 3 describes the quantum chemical based COSMO-SAC model to the liquid–liquid extraction process.
- Chapter 4 discusses a further application of the COSMO-SAC model in complex phase behavior such as vapor–liquid–liquid equilibria (VLLE).
- Chapter 5 covers a modification of the hydrogen bonding in the COSMO-SAC theory and applications to LLE systems.
- Chapter 6 defines the particle swarm optimization (PSO) technique that scales up the laboratory scale liquid–liquid extraction process to the industry scale multi-stage extraction process with optimum process parameters.
- Chapter 7 describes another optimization tool, the Cuckoo Search optimization technique, along with its application of the global optimization technique to liquid–liquid equilibria.

Learning Tools

Emphasis on Modelling

A distinctive feature of this book is an emphasis on a priori prediction of the properties of ILs and phase behavior of complex systems. Readers will acquire adequate knowledge about modelling of phase behavior from pseudo-algorithms given in this book.

State-of-the-Art Characterization Technique

Readers will gain knowledge of the analysis technique of liquid–liquid extraction data from nuclear magnetic resonance spectroscopy. A detailed analysis will give readers a ready-to-use database for composition analysis.

Predictive Database

The predictive models and experimental database reported in this book will act as a guide to researchers and engineers for scientific calculations. This will encourage them to work with similar systems at the expense of readily available models and databases.

MATLAB® is a registered trademark of The MathWorks, Inc. For product information, please contact:

The MathWorks, Inc.
3 Apple Hill Drive
Natick, MA 01760-2098 USA
Tel: 508-647-7000
Fax: 508-647-7001
E-mail: info@mathworks.com
Web: www.mathworks.com

About the Authors

Anand Bharti is currently a research scholar in the Department of Chemical Engineering, Indian Institute of Technology Guwahati, Guwahati, India.

Debashis Kundu is currently a research scholar in the Department of Chemical Engineering, Indian Institute of Technology Guwahati, Guwahati, India.

Dharamashi Rabari is an assistant professor at the School of Engineering and Applied Science, Ahmedabad University, Ahmedabad, India.

Tamal Banerjee is an associate professor in the Department of Chemical Engineering, Indian Institute of Technology Guwahati, Guwahati, India.

1

Introduction

Environment regulations and a strict legal framework motivate scientists and researchers to develop clean technologies for sustainable development. Researchers amplify their efforts to convert separation processes like extraction and distillation into clean technologies. The knowledge of thermodynamic properties like equilibrium data and activity coefficient at infinite dilution is required for adequate separation. Liquid–liquid equilibrium (LLE) data help in the selection of solvent for the selective separation of the solute component. In addition to this economic aspect, clean technology also considers the ecological aspect. Since the past few decades, eco-friendly solvents like ionic liquids (ILs) have been considered as a better replacement for conventional volatile solvents (Brennecke & Maginn, 2001; Welton, 1999). The rate of publication in the open literature has been rising at a greater rate for years. A total of 12,400 papers have been published since 2007 against 7,400 published between 1990 and March 2007. These data show the increase in new applications of ILs (Martins et al., 2014).

ILs are salts consisting of organic cations and inorganic or organic anions. Room temperature ionic liquids (RTILs) are liquids at room temperature. They have a bulky nonsymmetrical organic cation and an organic/inorganic anion. They are known as designer solvents due to the possibility of a large number of ILs (cation and anion combinations) which impart specific properties. These solvents are more attractive due to desirable special properties, such as wider liquid range, negligible vapor pressure at room temperature, tunable viscosity, low flammability, high heat capacity and favorable solvation properties for polar and nonpolar compounds (Anderson, Lin, Kuntzler, & Anseth, 2011; Mukherjee, Manna, Dinda, Ghosh, & Moulik, 2012; Neves, Carvalho, Freire, & Coutinho, 2011). These solvents are more stable at high temperatures and/or in the presence of chemicals and are suitable for the extraction of inorganic and organic compounds (Chapeaux, Simoni, Ronan, Stadtherr, & Brennecke, 2008; Garcia-Chavez, Garsia, Schuur, & de Haan, 2012; Simoni, Chapeaux, Brennecke, & Stadtherr, 2010; Wasserscheid & Keim, 2000; Welton, 1999). The physical properties of these solvents can be tuned by a suitable combination of cations and anions. Different types of RTILs are based on cations such as imidazolium, phosphonium, pyridinium, pyrrolidinium, ammonium and basionics. Salts are formed with anions possessing low nucleophilicity such as bis(trifluoromethylsulfonyl) imide, hexafluorophosphate, tetrafluoroborate, perfluoroalkylsulfonate, thiocynate, nitrate, acetate, ethyl sulphate and methane sulfonate.

The applications of ILs cover many research domains like catalysis (Tao, Zhuang, Cui, & Xu, 2014; Xiao, Su, Yue, & Wu, 2014; Ying et al., 2015; Zhuo et al., 2015), nanotechnology (Beier, Andanson, Mallat, Krumeich, & Baiker, 2012), azeotropic separation (Cai, Zhao, Wang, Wang, & Xiao, 2015), liquid–liquid extraction (García, García, Larriba, Torrecilla, & Rodríguez, 2011; García, Larriba, García, Torrecilla, & Rodríguez, 2011), metal extraction (Papaiconomou, Vite, Goujon, Leveque, & Billard, 2012), aromatic–aliphatic separation (Manohar, Banerjee, & Mohanty, 2013; Shah, Anantharaj, Banerjee, & Yadav, 2013), membrane science (Santos, Albo, & Irabien, 2014) and gas absorption (Shiflett, Niehaus, & Yokozeki, 2010, 2011). Today ILs are not just confined to laboratories; they have some applications at industrial scale as well, for example, BASF (BASIL), IFP (Difasol), PetroChina (Ionikylation), Linde (hydraulic ionic liquid compressor), Pionics (batteries) (Berthod, Ruiz-Ángel, & Carda-Broch, 2008; Marsh, Boxall, & Lichtenthaler, 2004; Plechkova & Seddon, 2008).

The database of ILs is increasing day by day, but still it is not possible to explore all ILs. Therefore, prediction or even correlation of their thermophysical properties is essential and useful for engineering applications. Considering the limited database reported in the literature, prediction or even correlation of their thermophysical properties is essential and useful for engineering applications.

Extraction and recovery of valuable chemicals and products are the prime concern of any chemical industry. One such promising extraction process is biofuels such as ethanol, propanol and butanol from fermentation broth. Biobutanol is typically produced via the acetone–butanol–ethanol fermentation process using renewable feedstock. Butanol has been identified as a superior biofuel with excellent fuel properties. Compared to ethanol and other fermentation-derived fuels, butanol offers several advantages as a biofuel such as higher energy content, lower volatility, lower hygroscopy and better miscibility with gasoline. Apart from its use as a biofuel, butanol also makes a suitable platform chemical for further processing to advanced biofuels such as butyl levulinate. Ethanol and easily mixed with gasoline in any proportion. Butanol can be obtained from a petrochemical route as well as biochemical route. The butanol produced by fermentation is very dilute. Different ILs have been considered as solvents based on density and hydrophobicity. Various hydrophobic ILs (Garcia-Chavez et al., 2012; Simoni et al., 2010) have shown better selectivity for butanol separation from aqueous solution and are more economical when compared to hydrophilic ILs for extraction of water. The composition analysis can be carried out by nuclear magnetic resonance (NMR) spectroscopy (Anantharaj & Banerjee, 2011; Potdar, Anantharaj, & Banerjee, 2012; Shah et al., 2013).

The last decade also saw considerable research in the fast pyrolysis process for the production of liquid fuel and chemicals from biomass. Fast pyrolysis of biomass produces 60–75 wt% of liquid bio-oil depending on the feedstock used. Due to high concentration of the value-added chemical compounds, production of chemicals from bio-oil has received considerable interest. Fractionation of bio-oil with water is the easiest method which transforms bio-oil into two fractions: an aqueous top phase enriched in carbohydrate-derived chemicals and an organic bottom phase containing lignin-containing fractions (Mohan, Pittman, & Steele, 2006; Vitasari, Meindersma, & de Haan, 2011). The aqueous phase derived from bio-oil is a good feed for the extraction of acetic acid, levoglucosan and sugar compounds. In place of conventional organic solvents, ILs can be used for the extraction of valuable chemicals from the aqueous phase derived from the bio-oil.

The experimental equilibrium data are usually correlated with thermodynamic models. Among them the calculation of activity coefficients is an integrated part in phase equilibrium calculations involving liquid phases. Activity coefficients are calculated via excess Gibbs free energy models like nonrandom two-liquid (NRTL) (Renon & Prausnitz, 1968), UNIversal QUAsiChemical (UNIQUAC) (Abrams & Prausnitz, 1975). Each of these models requires proper binary interaction parameters that can represent LLE for highly nonideal liquid mixtures. These parameters are usually estimated from the known experimental LLE data via optimization of a suitable objective function. The optimization problem can be either the least-squares objective function minimization or likelihood function maximization. In both cases the objective function is nonlinear and nonconvex in terms of optimization variables; this possesses several local minima/maxima/saddle points within the specified bounds of the variables.

Nature-inspired metaheuristic algorithms are becoming increasingly popular to solve global optimization problems. They work remarkably efficiently and have many advantages over traditional, deterministic methods. These algorithms are broadly classified into Evolutionary Algorithms, Physical Algorithms, Swarm Intelligence, Bio-inspired Algorithms and others (Nanda & Panda, 2014). Genetic Algorithm (GA), developed by John Holland and his collaborators, is one of the most widely used optimization algorithms in modern nonlinear optimization (Yang, 2010). Cuckoo Search (CS) developed by Yang and Deb (2009) is a population-based method which mimics the breeding behavior of certain cuckoo species and is one of the latest nature-inspired metaheuristic algorithms. Recent studies have shown that CS is potentially far more efficient than other algorithms in many applications. In this book, binary interaction parameters were estimated using the GA and CS algorithm and a comparison has been made between these two algorithms.

COSMO (conductor-like screening model) (Klamt, 1995) based models such as COSMO-SAC (conductor-like screening model segment activity coefficients) also determine liquid phase nonideality using molecular interactions derived from quantum chemical solvation calculation. COSMO-RS parameters are optimized against a large data set of experimental data and can be used for other types of systems without sacrificing accuracy. By this way, the model can be termed as a predictive model. In COSMO-based models, a molecule is moved from a vacuum to a perfect conductor and then to a real solvent. The molecules are regarded as a collection of surface segments and chemical potential of each segment is self-consistently determined from statistical mechanical calculation. The difference in segment activity coefficient between mixture and pure liquid gives the segment activity coefficients and activity coefficient of a molecule is obtained from summation over the segment activity coefficients. Thermodynamic properties are necessary for more complex phase behaviors, such as vapor–liquid–liquid equilibrium (VLLE). If the presence of heterogeneous liquid mixtures is not correctly accounted for the systems that exhibit a miscibility gap, there may be multiple solutions to vapor–liquid phases, which in turn may be a possible reason for multiple steady states in heterogeneous distillation. The condition at which VLLE occurs can be obtained from experimental measurements; however, it is often economical and time saving to predict computationally. The objectives of VLLE computations are to determine which of the three phases are present and to determine the composition of phases. In the original COSMO-SAC modelling (Lin & Sandler, 2002), the hydrogen donors or acceptors were considered when their charge density exceeds a certain threshold value (e.g. $|0.0084|$ e/Å^2), which leads to the step function change in probability. However for aqueous systems, this approach did not converge with the modified Rachford–Rice algorithm, which led to the failure in predicting the mole fractions of extract and raffinate phases. We have thus adopted a continuous probability distribution function (Hsieh, Sandler, & Lin, 2010; Lin, Chang, Wang, Goddard, & Sandler, 2004; Wang, Sandler, & Chen, 2007) of charge density so that for the acceptor and donor segments, higher the charge density, the greater the possibility of forming a hydrogen bond. The hydrogen bonding portion was obtained from the combination of the electronegative atom and hydrogen atom only.

Equation of state (EoS) have also assumed an important place in the prediction of phase equilibria of fluids and fluid mixtures. Using Wertheim's first-order thermodynamic perturbation theory, Chapman et al. (Chapman, Gubbins, Jackson, & Radosz, 1989, 1990) and Huang and Radosz (1990, 1991) developed the statistical associating fluid theory (SAFT) for pure fluids and mixtures containing associating fluids. Over the years, there have been numerous developments and modification of the SAFT EoS. In this contrast, the perturbed-chain statistical associating fluid theory (PC-SAFT) EoS developed by Gross and Sadowski (2001, 2002) has been successful for predicting the vapor–liquid and liquid–liquid equilibria of various systems based on only pure component parameters namely number of segments (m), segment diameter (σ), depth of potential (ε/k_B), association energy (ε^{AiBj}/k) and effective association volume (κ^{AiBj}). In Chapter 2, the PC-SAFT EoS has been used to predict the ternary LLE.

For transforming laboratory data into industrial application, the process optimization study is also necessary and useful. The binary interaction parameters generated by the NRTL model have been used for multi-stage extractor optimization. In the past, many popular stochastic algorithms such as GA (Goldberg, 1989), Simulated Annealing (Kirkpatrick, Gelatt, & Vecchi, 1983), particle swarm optimization (PSO) (Eberhart & Kennedy, 1995) and Differential Evolution (Storn & Price, 1997) have been investigated for optimization in science and engineering. PSO is an evolutionary algorithm based on social behavior of birds in swarm. The initial position and velocity of each particle are initiated randomly. During simulation, each particle in swarm (population) updates its position and velocity based on its experience as well as neighbors' experience within the search space. The PSO technique was employed for optimizing the flow rate and number of stages in a multi-stage extractor. A multi-stage extractor containing more than two components requires detailed design like temperature, pressure, flow rate and composition in each stage. These are achieved by solving material balance equations (M), phase equilibrium relation (E), mole fraction summation for each stage (S) and energy balance equations (H), better known as MESH equations. In this book, the traditional Isothermal Sum Rate method (Tsuboka & Katayama, 1976) has been considered for the stage-wise calculation. The optimum number of stages and solvent flow rate were obtained by minimizing the multi-stage extractor cost for extraction of butanol and ethanol from aqueous solution using ILs.

References

Abrams, D. S., & Prausnitz, J. M. (1975). Statistical thermodynamics of liquid mixtures: A new expression for the excess Gibbs energy of partly or completely miscible systems. *AIChE Journal, 21*(1), 116–128. doi:10.1002/aic.690210115.

Anantharaj, R., & Banerjee, T. (2011). Liquid–liquid equilibria for quaternary systems of imidazolium based ionic liquid + thiophene + pyridine + iso-octane at 298.15 K: Experiments and quantum chemical predictions. *Fluid Phase Equilibria, 312*, 20–30. doi:10.1016/j.fluid.2011.09.006.

Anderson, S. B., Lin, C. C., Kuntzler, D. V., & Anseth, K. S. (2011). The performance of human mesenchymal stem cells encapsulated in cell-degradable polymer-peptide hydrogels. *Biomaterials, 32*, 3564–3574.

Beier, M. J., Andanson, J.-M., Mallat, T., Krumeich, F., & Baiker, A. (2012). Ionic liquid-supported Pt nanoparticles as catalysts for enantioselective hydrogenation. *ACS Catalysis, 2*(3), 337–340. doi:10.1021/cs2006197.

Berthod, A., Ruiz-Ángel, M. J., & Carda-Broch, S. (2008). Ionic liquids in separation techniques. *Journal of Chromatography A, 1184*(1–2), 6–18. doi:10.1016/j.chroma.2007.11.109.

Brennecke, J. F., & Maginn, E. J. (2001). Ionic liquids: Innovative fluids for chemical processing. *AIChE Journal, 47*(11), 2384–2389. doi:10.1002/aic.690471102.

Cai, F., Zhao, M., Wang, Y., Wang, F., & Xiao, G. (2015). Phosphoric-based ionic liquids as solvents to separate the azeotropic mixture of ethanol and hexane. *The Journal of Chemical Thermodynamics, 81*, 177–183. doi:10.1016/j.jct.2014.09.019.

Chapeaux, A., Simoni, L. D., Ronan, T. S., Stadtherr, M. A., & Brennecke, J. F. (2008). Extraction of alcohols from water with 1-hexyl-3-methylimidazolium bis(trifluoromethylsulfonyl)imide. *Green Chemistry, 10*(12), 1301–1306. doi:10.1039/b807675h.

Chapman, W. G., Gubbins, K. E., Jackson, G., & Radosz, M. (1989). SAFT: Equation-of-state solution model for associating fluids. *Fluid Phase Equilibria, 52*, 31–38. doi:10.1016/0378-3812(89)80308-5.

Chapman, W. G., Gubbins, K. E., Jackson, G., & Radosz, M. (1990). New reference equation of state for associating liquids. *Industrial & Engineering Chemistry Research, 29*(8), 1709–1721. doi:10.1021/ie00104a021.

Eberhart, R., & Kennedy, J. (1995, October). A new optimizer using particle swarm theory. In *Proceedings of the sixth international symposium on micro machine and human science* (pp. 39–43) Vol. 1, Nagoya, Japan, IEEE Press, Piscataway, NJ. doi: 10.1109/MHS.1995.494215.

Fredenslund, A., Jones, R. L., & Prausnitz, J. M. (1975). Group-contribution estimation of activity coefficients in nonideal liquid mixtures. *AIChE Journal, 21*(6), 1086–1099. doi:10.1002/aic.690210607.

Garcia-Chavez, L. Y., Garsia, C. M., Schuur, B., & de Haan, A. B. (2012). Biobutanol recovery using nonfluorinated task-specific ionic liquids. *Industrial & Engineering Chemistry Research, 51*(24), 8293–8301. doi:10.1021/ie201855h.

García, S., García, J., Larriba, M., Torrecilla, J. S., & Rodríguez, F. (2011). Sulfonate-based ionic liquids in the liquid–liquid extraction of aromatic hydrocarbons. *Journal of Chemical & Engineering Data, 56*(7), 3188–3193. doi:10.1021/je200274h.

García, S., Larriba, M., García, J., Torrecilla, J. S., & Rodríguez, F. (2011). Liquid–liquid extraction of toluene from heptane using 1-alkyl-3-methylimidazolium Bis(trifluoromethylsulfonyl)imide ionic liquids. *Journal of Chemical & Engineering Data, 56*(1), 113–118. doi:10.1021/je100982h.

Goldberg, D. E. (1989). *Genetic Algorithms in component parameters Search, Optimization and Machine Learning*. Boston, MA: Addison-Wesley Longman Publishing Co.

Gross, J., & Sadowski, G. (2001). Perturbed-chain SAFT: An equation of state based on a perturbation theory for chain molecules. *Industrial & Engineering Chemistry Research, 40*(4), 1244–1260. doi:10.1021/ie0003887.

Gross, J., & Sadowski, G. (2002). Application of the perturbed-chain SAFT equation of state to associating systems. *Industrial & Engineering Chemistry Research, 41*(22), 5510–5515. doi:10.1021/ie010954d.

Hsieh, C.-M., Sandler, S. I., & Lin, S.-T. (2010). Improvements of COSMO-SAC for vapor–liquid and liquid–liquid equilibrium predictions. *Fluid Phase Equilibria, 297*(1), 90–97. doi:10.1016/j.fluid.2010.06.011.

Huang, S. H., & Radosz, M. (1990). Equation of state for small, large, polydisperse, and associating molecules. *Industrial & Engineering Chemistry Research, 29*(11), 2284–2294. doi:10.1021/ie00107a014.

Huang, S. H., & Radosz, M. (1991). Equation of state for small, large, polydisperse, and associating molecules: Extension to fluid mixtures. *Industrial & Engineering Chemistry Research, 30*(8), 1994–2005. doi:10.1021/ie00056a050.

Kirkpatrick, S., Gelatt, C. D., & Vecchi, M. P. (1983). Optimization by simulated annealing. *Science, 220*(4598), 671–680. doi:10.1126/science.220.4598.671.

Klamt, A. (1995). Conductor-like screening model for real solvents: A new approach to the quantitative calculation of solvation phenomena. *The Journal of Physical Chemistry, 99*(7), 2224–2235. doi:10.1021/j100007a062.

Lin, S.-T., Chang, J., Wang, S., Goddard, W. A., & Sandler, S. I. (2004). Prediction of vapor pressures and enthalpies of vaporization using a COSMO solvation model. *The Journal of Physical Chemistry A, 108*(36), 7429–7439. doi:10.1021/jp048813n.

Lin, S.-T., & Sandler, S. I. (2002). A priori phase equilibrium prediction from a segment contribution solvation model. *Industrial & Engineering Chemistry Research, 41*(5), 899–913. doi:10.1021/ie001047w.

Manohar, C. V., Banerjee, T., & Mohanty, K. (2013). Co-solvent effects for aromatic extraction with ionic liquids. *Journal of Molecular Liquids, 180*, 145–153. doi:10.1016/j.molliq.2013.01.019.

Marsh, K. N., Boxall, J. A., & Lichtenthaler, R. (2004). Room temperature ionic liquids and their mixtures—A review. *Fluid Phase Equilibria, 219*(1), 93–98. doi:10.1016/j.fluid.2004.02.003.

Martins, M. A. P., Frizzo, C. P., Tier, A. Z., Moreira, D. N., Zanatta, N., & Bonacorso, H. G. (2014). Update 1 of: Ionic liquids in heterocyclic synthesis. *Chemical Reviews, 114*(20), PR1–PR70. doi:10.1021/cr500106x.

Mohan, D., Pittman, C. U., & Steele, P. H. (2006). Pyrolysis of wood/biomass for bio-oil: A critical review. *Energy & Fuels, 20*(3), 848–889. doi:10.1021/ef0502397.

Mukherjee, I., Manna, K., Dinda, G., Ghosh, S., & Moulik, S. P. (2012). Shear- and temperature-dependent viscosity behavior of two phosphonium-based ionic liquids and surfactant triton X-100 and their biocidal activities. *Journal of Chemical & Engineering Data, 57*(5), 1376–1386. doi:10.1021/je200938k.

Nanda, S. J., & Panda, G. (2014). A survey on nature inspired metaheuristic algorithms for partitional clustering. *Swarm and Evolutionary Computation, 16*, 1–18. doi:10.1016/j.swevo.2013.11.003.

Neves, C. M. S. S., Carvalho, P. J., Freire, M. G., & Coutinho, J. A. P. (2011). Thermophysical properties of pure and water-saturated tetradecyltrihexylphosphonium-based ionic liquids. *The Journal of Chemical Thermodynamics, 43*(6), 948–957. doi:10.1016/j.jct.2011.01.016.

Papaiconomou, N., Vite, G., Goujon, N., Leveque, J.-M., & Billard, I. (2012). Efficient removal of gold complexes from water by precipitation or liquid–liquid extraction using ionic liquids. *Green Chemistry, 14*(7), 2050–2056. doi:10.1039/c2gc35222b.

Plechkova, N. V., & Seddon, K. R. (2008). Applications of ionic liquids in the chemical industry. *Chemical Society Reviews, 37*(1), 123–150. doi:10.1039/b006677j.

Potdar, S., Anantharaj, R., & Banerjee, T. (2012). Aromatic extraction using mixed ionic liquids: Experiments and COSMO-RS predictions. *Journal of Chemical & Engineering Data, 57*(4), 1026–1035. doi:10.1021/je200924e.

Renon, H., & Prausnitz, J. M. (1968). Local compositions in thermodynamic excess functions for liquid mixtures. *AIChE Journal, 14*(1), 135–144. doi:10.1002/aic.690140124.

Santos, E., Albo, J., & Irabien, A. (2014). Acetate based supported ionic liquid membranes (SILMs) for CO_2 separation: Influence of the temperature. *Journal of Membrane Science, 452*, 277–283. doi:10.1016/j.memsci.2013.10.024.

Shah, M. R., Anantharaj, R., Banerjee, T., & Yadav, G. D. (2013). Quaternary (liquid + liquid) equilibria for systems of imidazolium based ionic liquid + thiophene + pyridine + cyclohexane at 298.15 K: Experiments and quantum chemical predictions. *The Journal of Chemical Thermodynamics, 62*, 142–150. doi:10.1016/j.jct.2013.02.020.

Shiflett, M. B., Niehaus, A. M. S., & Yokozeki, A. (2010). Separation of CO_2 and H_2S using room-temperature ionic liquid [bmim][$MeSO_4$]. *Journal of Chemical & Engineering Data, 55*(11), 4785–4793. doi:10.1021/je1004005.

Shiflett, M. B., Niehaus, A. M. S., & Yokozeki, A. (2011). Separation of N_2O and CO_2 using room-temperature ionic liquid [bmim][BF4]. *The Journal of Physical Chemistry B, 115*(13), 3478–3487. doi:10.1021/jp107879s.

Simoni, L. D., Chapeaux, A., Brennecke, J. F., & Stadtherr, M. A. (2010). Extraction of biofuels and biofeedstocks from aqueous solutions using ionic liquids. *Computers & Chemical Engineering, 34*(9), 1406–1412. doi:10.1016/j.compchemeng.2010.02.020.

Storn, R., & Price, K. (1997). Differential evolution—A simple and efficient heuristic for global optimization over continuous spaces. *Journal of Global Optimization, 11*(4), 341–359. doi:10.1023/a:1008202821328.

Tao, F.-R., Zhuang, C., Cui, Y.-Z., & Xu, J. (2014). Dehydration of glucose into 5-hydroxymethylfurfural in SO_3H-functionalized ionic liquids. *Chinese Chemical Letters, 25*(5), 757–761. doi:10.1016/j.cclet.2014.01.044.

Tsuboka, T., & Katayama, T. (1976). General design algorithm based on pseudo-equilibrium concept for multistage multi-component liquid-liquid separation processes. *Journal of Chemical Engineering of Japan, 9*(1), 40–45. doi:10.1252/jcej.9.40.

Vitasari, C. R., Meindersma, G. W., & de Haan, A. B. (2011). Water extraction of pyrolysis oil: The first step for the recovery of renewable chemicals. *Bioresource Technology, 102*(14), 7204–7210. doi:10.1016/j.biortech.2011.04.079.

Wang, S., Sandler, S. I., & Chen, C.-C. (2007). Refinement of COSMO–SAC and the applications. *Industrial & Engineering Chemistry Research, 46*(22), 7275–7288. doi:10.1021/ie070465z.

Wasserscheid, P., & Keim, W. (2000). Ionic liquids—New "solutions" for transition metal catalysis. *Angewandte Chemie International Edition, 39*(21), 3772–3789. doi:10.1002/1521-3773(20001103)39:21<3772::aid-anie3772>3.0.co;2-5.

Welton, T. (1999). Room-temperature ionic liquids. Solvents for synthesis and catalysis. *Chemical Reviews, 99*(8), 2071–2084. doi:10.1021/cr980032t.

Xiao, L., Su, D., Yue, C., & Wu, W. (2014). Protic ionic liquids: A highly efficient catalyst for synthesis of cyclic carbonate from carbon dioxide and epoxides. *Journal of CO_2 Utilization, 6*, 1–6. doi:10.1016/j.jcou.2014.01.004.

Yang, X. S. (2010). *Engineering optimization: An introduction with metaheuristic applications*. Hoboken, NJ: John Wiley & Sons.

Ying, A., Li, Z., Ni, Y., Xu, S., Hou, H., & Hu, H. (2015). Novel multiple-acidic ionic liquids: Green and efficient catalysts for the synthesis of bis-indolylmethanes under solvent-free conditions. *Journal of Industrial and Engineering Chemistry, 24*, 127–131. doi:10.1016/j.jiec.2014.09.019.

Yang, X.-S., & Deb, S. (2009, December 9–11). Cuckoo search via Levy flights. In *Proceedings of world congress on nature & biologically inspired computing (NaBIC 2009)* (pp. 210–214), Coimbatore, India. New York: IEEE Publications.

Zhuo, K., Du, Q., Bai, G., Wang, C., Chen, Y., & Wang, J. (2015). Hydrolysis of cellulose catalyzed by novel acidic ionic liquids. *Carbohydrate Polymers, 115*, 49–53. doi:10.1016/j.carbpol.2014.08.078.

2

Liquid–Liquid Equilibria: Experiments, Correlation and Prediction

2.1 Introduction

In chemical, petrochemical, pharmaceutical and biochemical industries, valuable products are generally made by a combination of processes that include synthesis, reaction, separation and purification. Distillation, extraction, absorption, crystallization, adsorption and membrane-based processes are the most widely used unit operations for the separation and purification of various products. Distillation is by far the most widely used separation process that is based on isolating components from a liquid mixture based on the differences in their boiling points. But when components have close-boiling points or do not withstand a high temperature, distillation becomes ineffective. In these cases, liquid extraction is one of the main alternatives to consider, which utilizes chemical differences instead of boiling point differences to separate the components. Liquid–liquid extraction is a versatile unit operation which involves two immiscible liquid phases. The extract is the liquid phase which consists of solvent and extracted solute; while the raffinate is the solute lean phase or liquid phase from which solute has been removed (McCabe, Smith, & Harriott, 1993). Liquid–liquid equilibria (LLE) extraction processes have been successfully used in petrochemical, pharmaceutical and food industries. These have been in use since a long time in the oil industry on a large scale for removing aromatic compounds from gasoline or kerosene. These are also used for extracting aromatic compounds from lube oil stocks in order to produce lubricants and BTX (benzene, toluene, xylenes) aromatics (Wauquier, 2000). One of the most important applications of LLE in pharmaceuticals industry is in the recovery of penicillin (McCabe et al., 1993).

The existing liquid–liquid based extraction processes use conventional organic volatile solvents. Last few decades saw the development of ionic liquids (ILs) as a new class of solvents that have the potential to replace the conventional solvents in numerous chemical and industrial processes. There are some industrial processes that already use ILs such as the BASIL (Biphasic Acid Scavenging utilizing Ionic Liquids) process developed by the BASF and the Difasol process developed by the IFP, which is an improvement for the traditional Dimersol route (Plechkova & Seddon, 2008). Further various separation processes relevant for oil refineries have been studied using ILs as solvents, namely the desulfurization of petroleum-derived fuels such as diesel and gasoline (Arce, Francisco, & Soto, 2010; Francisco, Arce, & Soto, 2010; Gao, Guo, Xing, Zhao, & Liu, 2010; Kuhlmann, Haumann, Jess, Seeberger, & Wasserscheid, 2009; Li et al., 2009; Wu & Ondruschka, 2010), the selective separation of aromatic/aliphatic hydrocarbon mixtures (Arce, Earle, Katdare, Rodríguez, & Seddon, 2008; Arce, Earle, Rodríguez, Seddon, & Soto, 2009; Pereiro & Rodriguez, 2009) and extractive distillation (Lei, Li, & Chen, 2003).

In this chapter, we consider the feasibility of using ILs as solvents for two different extraction types: (1) extraction of bio-butanol from fermentation broth and (2) that of biochemicals (acetic acid and furfural) from bio-oil derived aqueous phase. Let us discuss them in detail with respect to the methodology, analysis and predictions. But before its detailed discussion, let us state the formulation for such an LLE process.

2.1.1 Extraction of Butanol from Fermentation Broth

Bio-butanol is typically produced via acetone–butanol–ethanol (ABE) fermentation of renewable feedstock using various strains of *Clostridium acetobutylicum* (Fischer, Klein-Marcuschamer, & Stephanopoulos, 2008) or *Clostridium beijerinckii* (Ha, Mai, & Koo, 2010) in anaerobic conditions resulting in the production of butanol, acetone and ethanol in a proportion of 6:3:1. Bio-butanol obtained via the ABE fermentation process is now considered as a potential biofuel as it has many advantages over other fermentation-derived fuels including ethanol. High concentration of butanol (>10 g/L) inhibits microbial cell growth during fermentation (Garcia-Chavez, Garsia, Schuur, & de Haan, 2012); however, its removal reduces the effect of product inhibition and enables the conversion of the concentrated feed leading to a high productivity. Accordingly, in situ recovery of butanol from fermentation broth has gained considerable attention. Several techniques such as stripping, adsorption, liquid–liquid extraction, pervaporation and membrane solvent extraction have been investigated for removing butanol from a fermentation broth. Among these methods, liquid–liquid extraction has shown advantages over the others (Ha et al., 2010). Liquid–liquid extraction can be performed with high selectivity and is possible to carry out inside a fermenter. Several studies have been done for the extraction of alcohols from aqueous solutions using ILs (Chapeaux, Simoni, Ronan, Stadtherr, & Brennecke, 2008; Garcia-Chavez et al., 2012; Ha et al., 2010; Simoni, Chapeaux, Brennecke, & Stadtherr, 2010).

Here, we consider the feasibility of using low-density (lighter than water) ILs as solvents for extracting butanol from aqueous solutions. Low-density ILs are beneficial as they reduce pumping charges resulting in the reduction of the operating cost. Liquid–liquid extraction of butanol from water is demonstrated using three different phosphonium-based ILs, namely trihexyl (tetradecyl) phosphonium bis (2,4,4-trimethylpentyl) phosphinate ([TDTHP][Phosph]), trihexyl (tetradecyl) phosphonium dicyanamide ([TDTHP][DCA]) and trihexyl (tetradecyl) phosphonium decanoate ([TDTHP][DEC]). Further, two quaternary systems, that is, [TDTHP][Phosph] + ethanol + butanol + water and [TDTHP][DCA] + ethanol + butanol + water, were also examined to study the effect of ethanol on butanol extraction. This is necessary as ethanol and butanol coexist in an ABE fermenter.

2.1.2 Extraction of Biochemicals (Acetic Acid and Furfural) from Aqueous Phase

The last decade saw considerable research in fast pyrolysis process for the production of liquid fuel and chemicals from biomass. Fast pyrolysis of biomass produces 60–75 wt% of liquid bio-oil, 15–25 wt% of solid char and 10–20 wt% of noncondensable gases depending on the feedstock used (Bridgwater, 2003, 2012; Mohan, Pittman, & Steele, 2006). Several chemicals have been identified in bio-oil, of which the ones most abundant and of interest are glycolaldehyde (0.9–13 wt%), acetic acid (0.5–12 wt%), formic acid (0.3–9.1 wt%), acetol (0.7–7.4 wt%), furfural alcohol (0.1–5.2 wt%) and furfural (0.1–1.1 wt%; Diebold, 2000). Due to high concentration of value-added chemical compounds, the production of chemicals from bio-oil has received considerable interest. Fractionation of bio-oil with water is the easiest method which transforms bio-oil in to two fractions: an aqueous top phase enriched in carbohydrate-derived chemicals and an organic bottom phase consisting lignin-containing fractions. Both phases can be further processed separately to extract value-added chemicals (Mohan et al., 2006; Vitasari, Meindersma, & de Haan, 2011). The aqueous phase derived from bio-oil is a good feed for the extraction of acetic acid, levoglucosan and sugar compounds. Several studies have been done for the extraction of acetic acid (Mahfud, van Geel, Venderbosch, & Heeres, 2008; Rasendra, Girisuta, Van deBovenkamp, Winkleman, Leijenhorst, Venderbosch, Windt, Meier, & Heeres, 2011), levoglucosan (Bennett, Helle, & Duff, 2009), sugar and sugar derivatives (Hu, Mourant, Gunawan, Wu, Wang, Lievens, & Li, 2012) using conventional solvents. Only few studies have been carried out for the extraction of acetic acid (Matsumoto, Mochiduki, Fukunishi, & Kondo, 2004; Yu, Li, & Liu, 2006) and furfural (Pei, Wu, Wang, & Fan, 2008) from aqueous solution using ILs. Therefore, in this formulation the feasibility of using ILs to extract two of the valuable chemicals, namely acetic acid and furfural, from the aqueous phase will be studied. Hence, LLE data will be measured for the following two systems: 1-butyl-3-methylimidazolium bis (trifluoromethylsulfonyl) imide ([BMIM][TF_2N]) + acetic acid + water and [BMIM][TF_2N] + furfural + water.

2.1.3 Predictive Models

Among the prediction insights, activity coefficient models such as non-random two-liquid (NRTL; Renon & Prausnitz, 1968) and UNIversal QUAsiChemical (UNIQUAC; Abrams & Prausnitz, 1975) are usually used to predict LLE.

Banerjee, Singh, Sahoo and Khanna (2005) have correlated LLE data for seven IL-based ternary systems by the UNIQUAC model and have found good agreement with experimental data. Aznar (2007) correlated LLE data for 24 IL-based ternary systems by the NRTL model and found good agreement between experimental and calculated compositions with global root mean square (RMS) deviations of 1.4%. Santiago, Santos and Aznar (2009) correlated LLE data of 50 ternary systems with 12 different ILs comprising 408 experimental tie lines of the UNIQUAC model. The results were very satisfactory with deviations of about 1.75%. Therefore, in this chapter, the NRTL and the UNIQUAC models were used to correlate the experimental tie-line data.

The above-mentioned models require proper binary interaction parameters. These parameters are usually estimated from the known experimental LLE data via the optimization of a suitable objective function. The objective function is highly nonconvex having multiple local optima within the specified bounds of the variables. Therefore, it is necessary to apply a technique that results in the global optimization of the variables. Genetic algorithm (GA), developed by John Holland and his collaborators, is one of the most widely used optimization algorithms in modern nonlinear optimization (Yang, 2010). In this chapter, GA has been used for the estimation of binary interaction parameters of the NRTL and the UNIQUAC models.

The ability to predict phase equilibria is essential for the simulation of new chemical processes and thus for process development. The predictive models should possess a sufficient accuracy over a wide range of conditions using a minimum of easily accessible parameters. Activity coefficient models, NRTL and UNIQUAC, are thus correlative in nature. These models are not able to predict the phase equilibria without the experimental LLE data.

Perturbed-chain statistical associating fluid theory (PC-SAFT) equation of state (EoS) developed by Gross and Sadowski (2001, 2002) is another predictive model which has been successful for predicting the vapor–liquid and liquid–liquid equilibria of various systems based on only pure component parameters, namely number of segments (m), segment diameter (σ), depth of potential (ε/k_B), association energy (ε^{AiBj}/k) and effective association volume (κ^{AiBj}). Therefore, in this chapter PC-SAFT EoS has been used to predict the ternary LLE. We now discuss the LLE method and its analysis in detail before we move on to discussing the two formulations in detail.

2.2 Extraction Methodology

In each LLE experiment a known quantity of components were added in a 15 mL size glass vial. Components were added in such a composition that two different layers formed. The glass vials were sealed using parafilm to avoid any evaporation loss. These sealed glass vials were then kept in a thermostatic shaker bath which was operated at 150–200 RPM and 298.15 K (uncertainty of ±0.01 K). The samples were stirred for 3–8 h at isothermal condition and kept undisturbed for 12–20 h to ensure the equilibrium. Two clear phases appeared, that is, the water-rich phase and the IL-rich phase. Then, samples from each phase were collected using syringes for compositional analysis. Feed compositions which were homogenous after equilibrium were rejected as their location was outside the binodal curve.

2.3 Characterization Techniques

Techniques such as ultraviolet–visible spectroscopy (UV–Vis), gas chromatography–mass spectrometry (GC–MS) and ^{1}H-nuclear magnetic resonance (NMR) spectroscopy are conventionally used for the compositional analysis. For composition determination using UV–Vis spectroscopy, a calibration curve is required for each component. For the preparation of the calibration curve, the absorbance of a series of solutions of known concentrations are measured and plotted against their corresponding concentrations to get a linear plot which follows the Beer–Lambert Law. However, these calibration charts become a tedious process for multi-component systems, especially for the system which has three components. The selection of solvent is also important as it should be a near-perfect material offering transparency over the UV–Vis range. Similarly, GC–MS is also widely used for the characterization of compounds. In GC–MS, the sample is injected into the GC inlet where it is vaporized and swept onto a chromatographic column by the carrier gas. Normal boiling temperatures of ILs are inherently high; hence, it becomes difficult to vaporize at injection. Therefore, modification is performed in the GC column by inserting a guard column that retains the IL and restricts it from entering the main GC column. In comparison to UV–Vis spectroscopy and GC–MS, ^{1}H-NMR spectroscopy requires less calibration data and consumes less time. Nowadays, NMR spectroscopy is widely used for determining LLE compositions (Arce, Earle, Katdare, Rodríguez, & Seddon, 2007; Arce, Rodríguez, & Soto, 2006; Pilli, Banerjee, & Mohanty, 2014).

^{1}H-NMR spectrum provides the information regarding the number of different types of hydrogen present in the molecule as well as the electronic environment of the different types of hydrogen. Each group of chemically equivalent hydrogen gives rise to a unique peak in the NMR spectrum and different groups provide different chemical shift. The number of hydrogen atoms corresponding to each peak is identified by the integration (relative area) of the peak for each group. In other words, the area under each peak is proportional to the number of hydrogen atoms generating that peak and this is used for the determination of the composition of each component present in the mixture. In our work, ^{1}H-NMR spectra were recorded by 400 MHz (Varian) and 600 MHz (Bruker) NMR spectrometers. NMR spectra are recorded in solution and it is assumed that solvent protons do not interfere with the compound spectra. Therefore, deuterated solvents are used for NMR. In this work, two solvents, namely deuterated dimethyl sulfoxide (DMSO-d_6), $(CD_3)_2SO$ and deuterated chloroform, $CDCl_3$, were used, because the raffinate phase consisting water is not miscible with $CDCl_3$. Due to this reason, DMSO-d6 was used as the NMR solvent while recording the spectrum of the raffinate phase.

A constant sample of 0.07–0.10 mL from each phase was collected and mixed with 0.5 mL of the NMR solvent in NMR tubes (thrift grade, Sigma–Aldrich). The closed top end of the tube was covered with parafilm tape to avoid any evaporation losses. The tubes were then placed in NMR spectrophotometer for ^1H analysis. The reference peak for the NMR solvent was also recorded (2.5 ppm for DMSO-d6 and 7.26 for $CDCl_3$). Figure 2.1 shows the typical spectra of the extract phase of [BMIM][TF_2N]–acetic acid–water system. The methyl group (-CH_3-) of acetic acid (CH_3COOH) showed a peak at approximately 1.9 ppm which was considered for the quantification of acetic acid in both the phases. IL [BMIM][TF_2N] ($CH_3CH_2CH_2CH_2(C_3N_2H_3)CH_3$) showed eight different peaks: three due to H-atoms attached to imidazolium ring, one due to

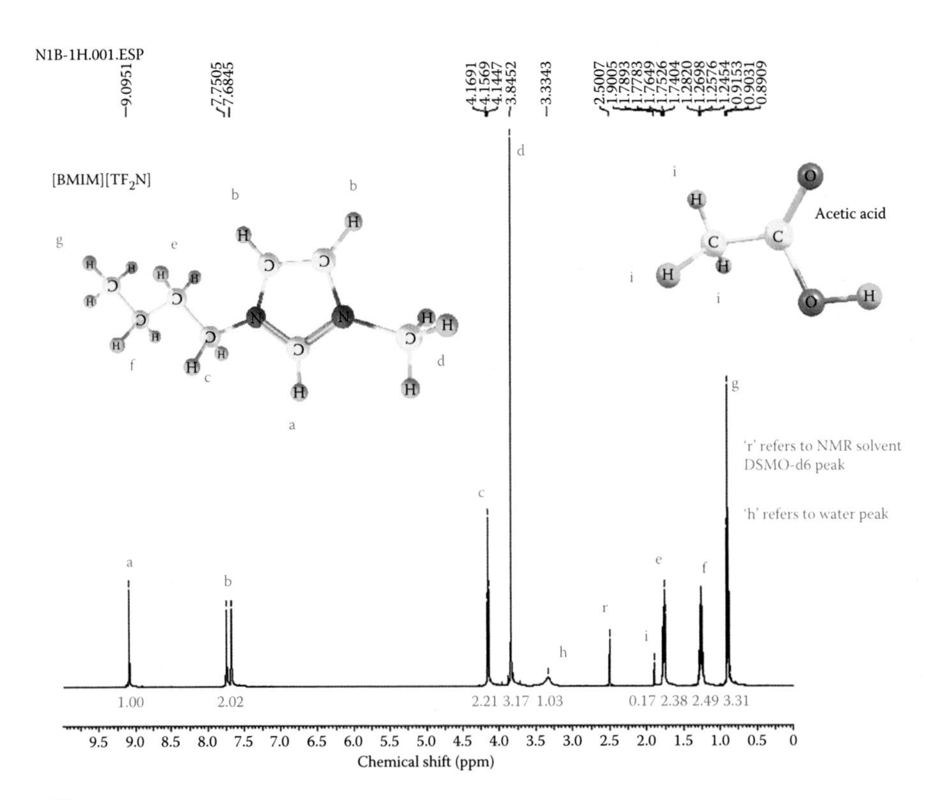

FIGURE 2.1
^1H-NMR spectra for the extract phase of [BMIM][TF_2N]–acetic acid–water system at $T = 298.15$ K and $p = 1$ atm.

3-methyl ($-CH_3-$) group, three due to $-CH_2-$ group of 1-butyl and one due to $-CH_3-$ group of 1-butyl. The peak at approximately 9.0 ppm due to a single H-atom was used for the quantification of IL in both the phases. The water in DMSO-d6 gave a peak at approximately 3.3–4.0 ppm. Similarly, other ^{1}H-NMR spectra were used for the identification of the peaks of various compounds for the test cases of the formulated systems studied in this work. The concentration of each component in each phase was then calculated by Equation 2.1:

$$x_i = \frac{H_i}{\sum_{i=1}^{3} H_i} \tag{2.1}$$

where:

H_i denotes the peak area of single hydrogen of component '*i*'
x_i the mole fraction of component '*i*'

The extraction effectiveness of a solvent is generally expressed in terms of distribution coefficient (β) and selectivity (*S*). Distribution coefficient (β) and selectivity (*S*) are defined as

$$\beta = \frac{x_{\text{solute}}^{E}}{x_{\text{solute}}^{R}} \tag{2.2}$$

$$S = \frac{\left(x_{\text{solute}}^{E} / x_{\text{solute}}^{R}\right)}{\left(x_{\text{w}}^{E} / x_{\text{w}}^{R}\right)} \tag{2.3}$$

where x_{solute} and x_{w} are the mole fractions of solute (i.e. alcohol, acetic acid or furfural) and water, respectively. Superscripts *E* and *R* indicate the extract and the raffinate phases, respectively.

2.4 UNIQUAC and Nonrandom Two-Liquid Model Equations

Gibb's free energy models namely UNIQUAC and NRTL were used to correlate experimental tie-line data. In the UNIQUAC model, the activity coefficient, γ_i, of component '*i*' in the multi-component system is given by

$$\ln\gamma_i = \ln\left(\frac{\Phi_i}{x_i}\right) + \frac{z}{2} q_i \ln\left(\frac{\theta_i}{\Phi_i}\right) + l_i - \frac{\Phi_i}{x_i}\sum_{j=1}^{c} x_j l_j + q\left(1 - \ln\sum_{j=1}^{c}\theta_j\tau_{ji} - \sum_{j=1}^{c}\frac{\theta_j\tau_{ij}}{\sum_{k=1}^{c}\theta_k\tau_{kj}}\right) \tag{2.4}$$

$$\text{with } \tau_{ij} = \exp\left(-\frac{A_{ij}}{T}\right), \quad \theta_i = \frac{q_i x_i}{\sum_k q_k x_k}, \quad \Phi_i = \frac{r_i x_i}{\sum_k r_k x_k}, \quad l_i = \frac{z}{2}(r_k - q_k) + 1 - r_k$$

where:

z is lattice coordination number ($z = 10$)

r_i and q_i are, respectively, the volume and the surface area of the pure component i

x_i is the mole fraction of component i

τ_{ij} is the interaction parameter between components i and j

In the NRTL model, the activity coefficient, γ_i, of component 'i' in the ternary system is given by

$$\ln \gamma_i = \frac{\sum_{j=1}^{c} \tau_{ji} G_{ji} x_j}{\sum_{l=1}^{c} G_{li} x_l} + \sum_{j=1}^{c} \frac{x_j G_{ij}}{\sum_{l=1}^{c} G_{lj} x_l} \left[\tau_{ij} - \frac{\sum_{r=1}^{c} x_r \tau_{rj} G_{rj}}{\sum_{l=1}^{c} G_{lj} x_l} \right] \tag{2.5}$$

with

$$G_{ji} = \exp(-\alpha_{ji}\tau_{ji}), \quad \tau_{ji} = \frac{g_{ji} - g_{ii}}{RT}, \quad \alpha_{ji} = \alpha_{ij}$$

where:

g is an energy parameter characterizing the interaction of species i and j

x_i is the mole fraction of component i

α is the non-randomness of the parameter.

Although α can be treated as an adjustable parameter, in this study α was set equal to 0.2 which is a value prescribed for hydrocarbons.

The polarizable continuum model (PCM; Banerjee et al., 2005) with the GEPOL algorithm was used to predict the volume parameter (r) and the surface area parameter (q) in the UNIQUAC model. For the estimation of r and q, Equations 2.6 and 2.7 have been used as

$$r_{\text{pred}} = \frac{\left(V^{\text{PCM}} \text{in}^3\right)\left(1 \times 10^{-8} \text{cm}\right) N_{av}}{V_{ws}} \tag{2.6}$$

$$q_{\text{pred}} = \frac{\left(A^{\text{PCM}} \text{in}^2\right)\left(1 \times 10^{-8} \text{cm}\right)^2 N_{av}}{A_{ws}} \tag{2.7}$$

where:

N_{av} is the Avagrado's number

V_{ws} is standard segment volume (i.e. 15.17 cm^3/mol)

A_{ws} is the area (2.5×10^9 cm^2/mol).

TABLE 2.1

UNIQUAC Volume and Surface Area Parameters for Different Components

Sr. No.	Compounds	Volume Parameter (r)	Surface Area Parameter (q)
1	[TDTHP][Phosph]	9.834	6.258
2	[TDTHP][DCA]	8.37	5.81
3	[TDTHP][DEC]	8.77	5.96
4	[BMIM][TF_2N]	11.964	9.753
5	Acetic acid	2.202	2.072
6	Furfural	3.168	2.481
7	Water	0.92	1.4
8	Ethanol	2.11	1.97
9	1-Propanol	3.2499	3.128
10	1-Butanol	3.9243	3.668

The output file of PCM contains the overall surface (A^{PCM}) and the overall volume (V^{PCM}). The parameters' values are reported in Table 2.1.

2.5 Genetic Algorithm for Prediction of Model Parameters

The thermodynamic equilibrium condition for multi-component liquid–liquid system can be described by the following expression:

$$\gamma_i^I x_i^I = \gamma_i^{II} x_i^{II} \qquad (i = 1, 2, 3, ...) \tag{2.8}$$

where γ_i, the activity coefficient of component i in a phase (I or II), is predicted using the NRTL/UNIQUAC model. x_i^I and x_i^{II} represent the mole fraction of component i in phases I and II, respectively.

The compositions of the extract and the raffinate phases are calculated using a flash algorithm as described by the modified Rachford–Rice algorithm (Seader & Henley, 2006). The optimum binary interaction parameters are those which minimize the difference between the experimental and the calculated compositions, and are given by the relation

$$F_{obj} = \sum_{k=1}^{m} \sum_{i=1}^{c} \sum_{l=I}^{II} \left(x_{ik}^{l} - \hat{x}_{ik}^{l} \right)^2 \tag{2.9}$$

Equation 2.9 is highly nonconvex in nature and is difficult to solve. This necessitates the requirement of a non-traditional optimization tool such as the GA. The root-mean-square deviation (RMSD) values, which provide a measure of the accuracy of the correlations, were calculated according to the following expression:

$$\text{RMSD} = \left(\frac{F_{\text{obj}}}{2mc}\right)^{1/2} = \left[\sum_{k=1}^{m}\sum_{i=1}^{c}\sum_{l=I}^{II}\frac{\left(x_{ik}^{l} - \hat{x}_{ik}^{l}\right)^2}{2mc}\right]^{1/2} \tag{2.10}$$

where:
- m is the number of tie lines
- c is the number of components

x_{ik}^{l} and $\hat{x}_{ik}^{l}$ are the experimental and predicted values of mole fraction for component i for the kth tie line in phase l, respectively.

Figure 2.2 shows the flow diagram of the total algorithm used in this work for the calculation of binary interaction parameters.

GA is one of the most widely used evolutionary optimization algorithm in modern nonlinear optimization. The GA, developed by John Holland and his collaborators, is based on Charles Darwin's theory of evolution and natural selection that mimics biological evolution (Goldberg, 1989; Deb, 2001 & Yang, 2014). Compared to the traditional derivative-based optimization algorithms GA is a population-based optimization algorithm and therefore, GA explores the search space with a population of solutions instead of a single solution. Thus, it is likely that the expected GA solution may be a global solution. The derivative-based algorithm generates a new point by a deterministic computation, whereas GA creates a new population by probabilistic rules.

The GA repeatedly modifies a population. At each step, the GA selects individuals at random from the current population to be parents and uses them to produce the children for the next generation. Over successive generations, the population 'evolves' towards an optimal solution. The GAs have three main genetic operators to create the next generation from the current population: (1) the *selection operator* which selects the individuals, called parents, that contribute to the population at the next generation, (2) *the crossover operator* which combines the two parents to form children for the next generation and (3) *the mutation operator* which applies random changes to individual parents to form children. The stochastic nature of crossover and mutation make GA explore the search space more effectively.

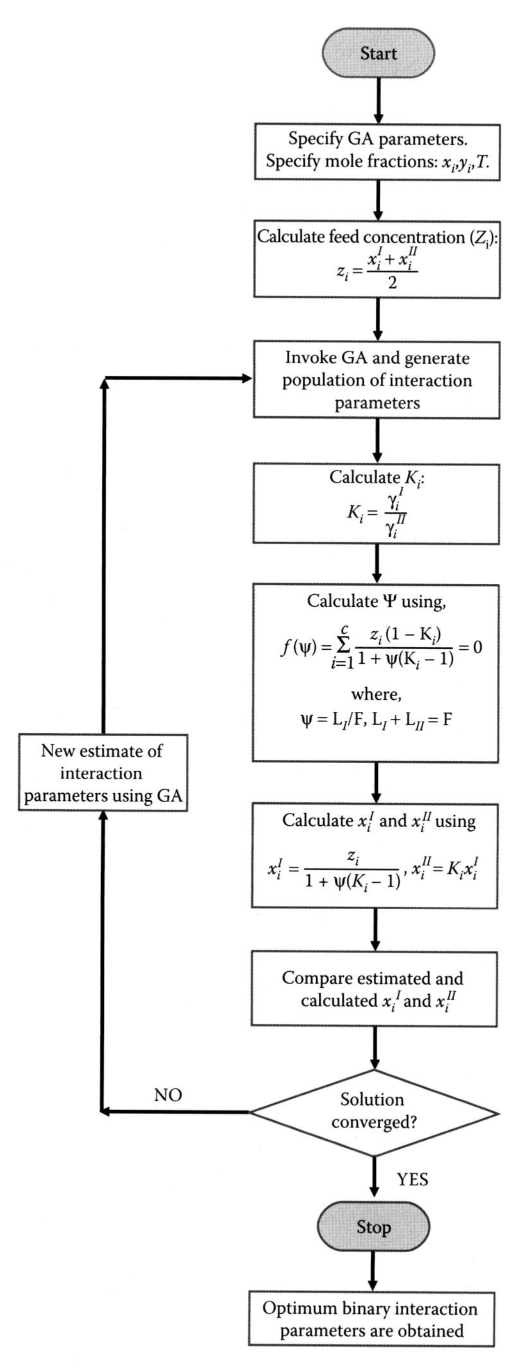

FIGURE 2.2
Flow diagram of the flash algorithm used for LLE modelling.

The steps of the algorithm are as follows:

Step 1: Initialization of GA. The algorithm starts with defining the fitness function or the objective function which is to be optimized. The following parameters namely the population size with lower and upper bounds of the variables are specified.

Step 2: Generation of initial population and fitness evaluation. The initial population is randomly generated between lower and upper bounds and the fitness value of each member of the population is computed.

Step 3: Generation of new population. At each step, the algorithm uses the individuals in the current generation to create the next generation, for which the algorithm performs the following steps:

Selection: Individuals of the current population, called parents, are selected based on their fitness. Some of the individuals in the current population that have best fitness are chosen as elite and these elite individuals are passed to the next generation directly.

Crossover: The algorithm creates members of the new population or children by combining pairs of parents in the current population. Crossover enables the algorithm to extract the best genes from different individuals and recombine them into potentially superior children.

Mutation: The algorithm creates mutation in children by randomly changing the genes of individual parents. The mutation operation prevents GA from converging to a local minimum and also introduces new possible solutions into the population.

Step 4: Fitness evaluation. The fitness evaluation of each member of the new population is computed.

Step 5: Stopping criterion. The algorithm stops when one of the stopping criteria is met such as (a) maximum number of generations, (b) the value of the fitness function for the best individual in the current population and (c) average relative change in the fitness function value over successive generations.

The flow diagram of the GA is shown in Figure 2.3. In this work, the GA toolbox as available in MATLAB® 7.10.0 (R2010a) has been used for all the LLE calculations.

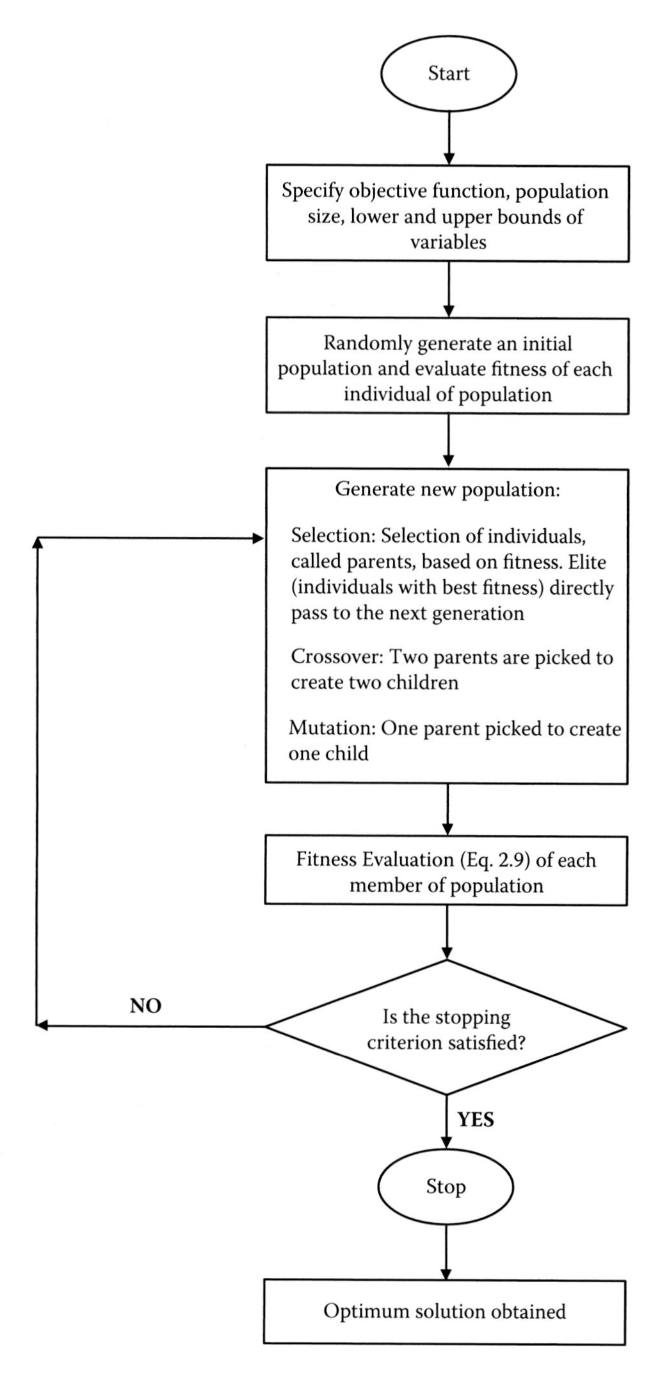

FIGURE 2.3
Flow diagram of GA.

2.6 Ternary and Quaternary LLE Systems

2.6.1 Ternary LLE Systems*

Distribution coefficient gives a measure of the affinity of the solute for the two phases. The distribution coefficient for a solute should be large enough so that a lower solvent (S)/feed (F) ratio is required. Further, the selectivity parameter must be greater than unity for a useful extraction operation. As discussed before, the results from the two problem formulations will be discussed. We shall start with the first formulated section namely 'Extraction of Butanol from Fermentation Broth' which is discussed in Section 2.1.1.

Experimental LLE ternary diagram for systems [TDTHP][Phosph] + 1-butanol + water, [TDTHP][DCA] + 1-butanol + water and [TDTHP][DEC] + 1-butanol + water are shown in Figures 2.4 through 2.9.

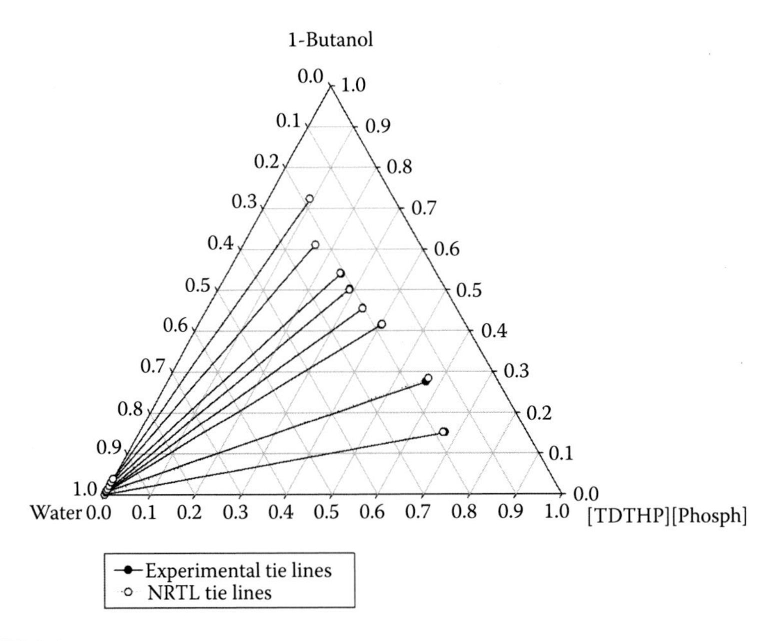

FIGURE 2.4
Experimental and NRTL-predicted tie lines for the ternary system 1-butanol/water/[TDTHP][Phosph] at $T = 298.15$ K and $p = 1$ atm.

* Section 2.6.1 reprinted (adapted) from D. Rabari, T. Banerjee, Biobutanol and n-propanol recovery using a low-density phosphonium-based ionic liquid at $T = 298.15$ K and $p = 1$ atm, *Fluid Phase Equilibria* 355, 26–33, 2013. Copyright 2013, with permission from Elsevier. D. Rabari, T. Banerjee, Experimental and theoretical studies on the effectiveness of phosphonium-based ionic liquids for butanol removal at $T = 298.15$ K and $p = 1$ atm, *Ind. Eng. Chem. Res.* 53, 18935–18942, 2014. Copyright 2014 American Chemical Society. A. Bharti, T. Banerjee, Enhancement of bio-oil-derived chemicals in aqueous phase using ionic liquids: experimental and COSMO-SAC predictions using a modified hydrogen bonding expression. *Fluid Phase Equilibria*. 400, 27–37, 2015. Copyright 2015, with permission from Elsevier.

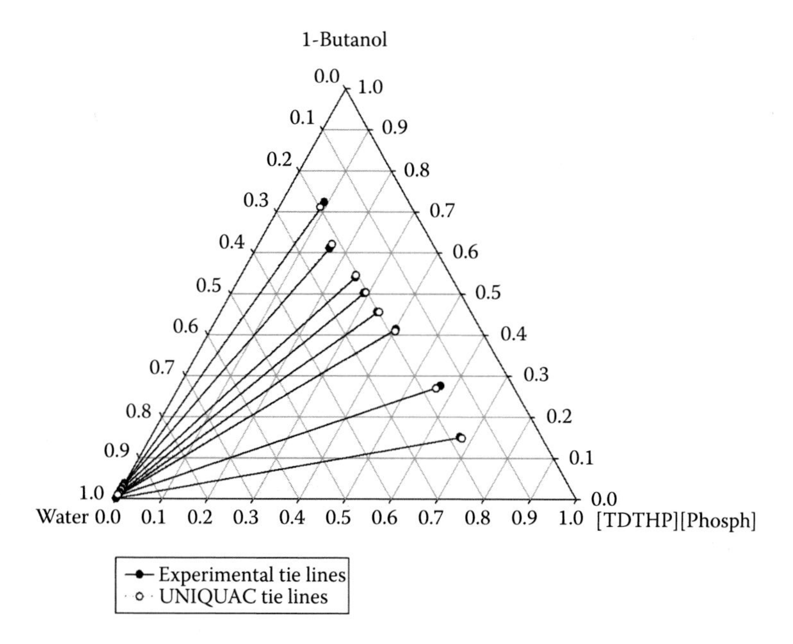

FIGURE 2.5
Experimental and UNIQUAC-predicted tie lines for the ternary system 1-butanol/water/[TDTHP][Phosph] at $T = 298.15$ K and $p = 1$ atm.

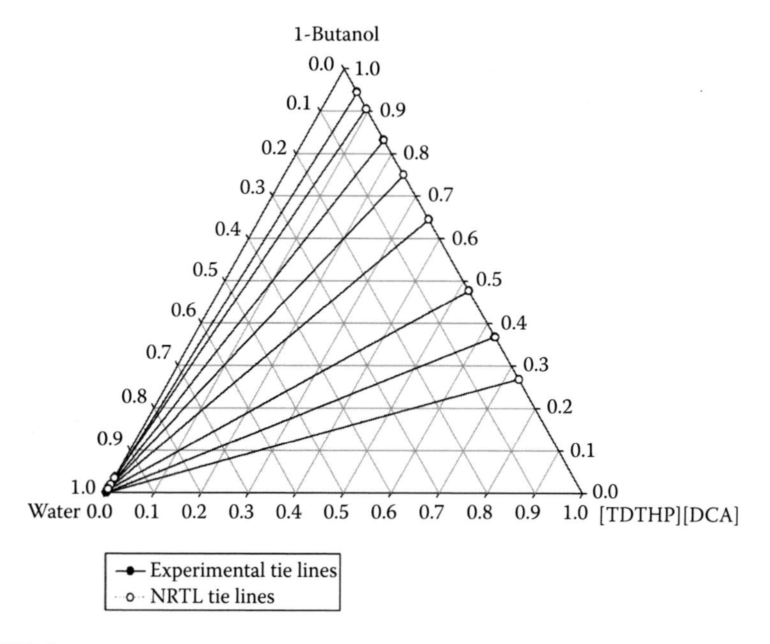

FIGURE 2.6
Experimental and NRTL tie lines for the ternary system [TDTHP][DCA]/1-butanol/water at $T = 298.15$ K and $p = 1$ atm.

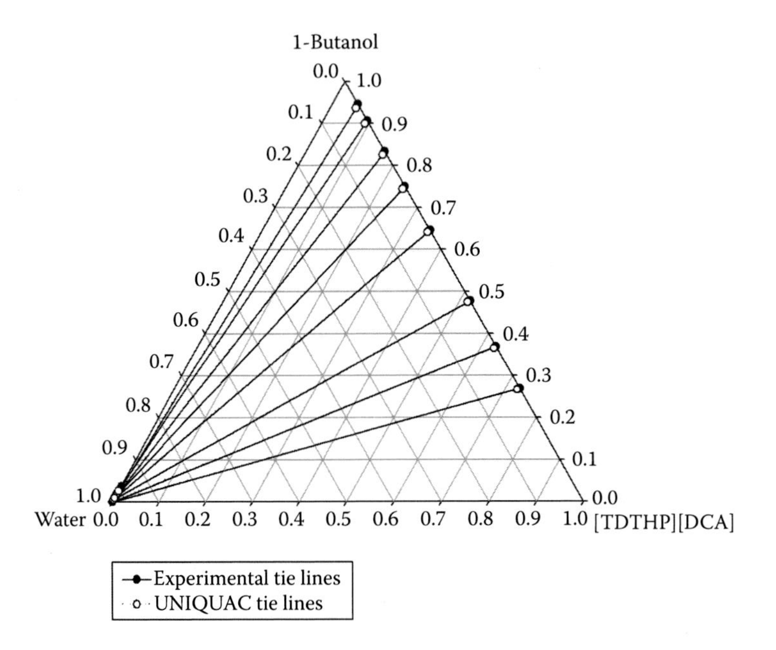

FIGURE 2.7
Experimental and UNIQUAC tie lines for the ternary system [TDTHP][DCA]/1-butanol/water at $T = 298.15$ K and $p = 1$ atm.

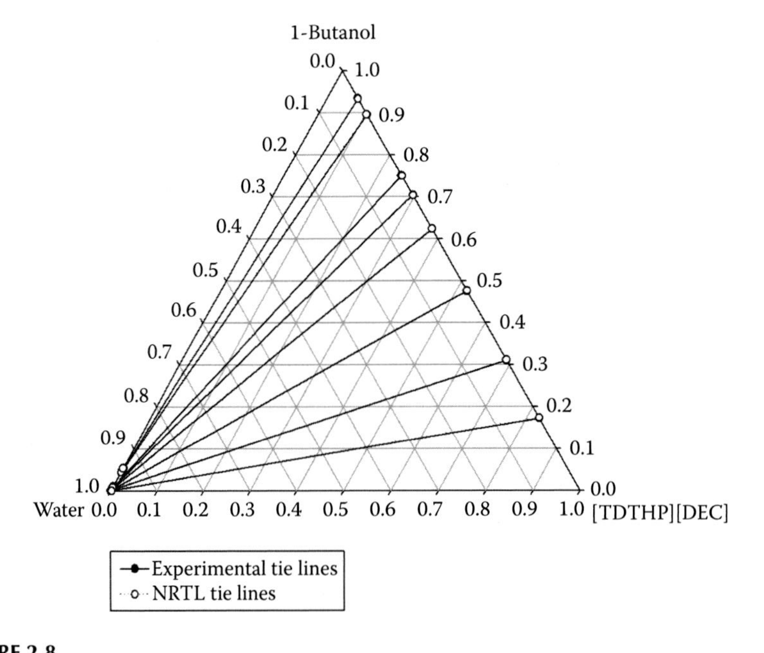

FIGURE 2.8
Experimental and NRTL tie lines for the ternary system [TDTHP][DEC]/1-butanol/water at $T = 298.15$ K and $p = 1$ atm.

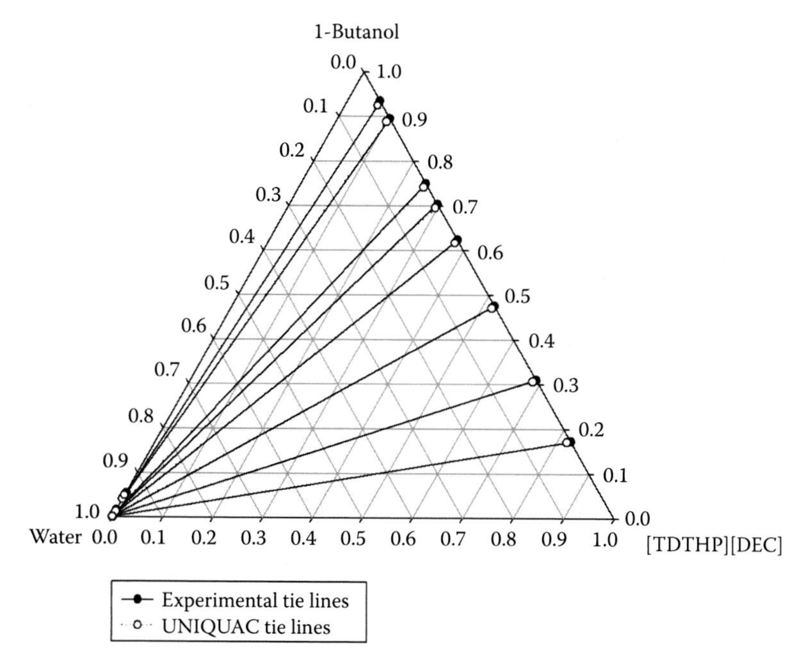

FIGURE 2.9
Experimental and UNIQUAC tie lines for the ternary system [TDTHP][DEC]/1-butanol/water at $T = 298.15$ K and $p = 1$ atm.

[TDTHP] [Phosph] shows the distribution coefficient for butanol extraction to be approximately 19–59. The distribution coefficient varied in the range of 25–95 and 17–173 for ILs [TDTHP][DCA] and [TDTHP][DEC], respectively. The positive sloping of the tie lines indicates that butanol favorably partitions into the IL phase. These values are much higher than 0.7–2.2 as obtained by Ha et al. (2010) for different imidazolium ILs containing tetrafluoroborate, trifluoromethanesulfonate, hexafluorophosphate and bis (trifluoromethylsulfonyl) imide anions. The distribution coefficient for [TDTHP][DCA] is twice as compared to the ILs reported by Nann, Held, and Sadowski (2013), that is, 1-decyl-3-methylimidazolium tetracyanoborate ([Im10.1][TCB]), 4-decyl-4-methylmorpholinium tetracyanoborate ([Mo10.1][TCB]), 1-decyl-3-methylimidazolium bis (trifluoromethylsulfonyl) ([Im10.1][Tf_2N]) and 4-decyl-4-methylmorpholinium bis-(trifluoromethylsulfonyl)imide ([Mo10.1][Tf_2N]). Further, it was also observed that the distribution coefficients for [TDTHP][Phosph] and [TDTHP][DEC] are, respectively, 12 and 7 times higher as compared to the same ILs (Nann et al., 2013). Thus, we can conclude that [TDTHP][Phosph] and [TDTHP][DEC] more favorably extract butanol from aqueous streams. The distribution coefficients for the three systems are again compared with literature data (Kubiczek & Kamiński, 2013) in Figure 2.10. It should be noted that the ternary diagram is wider than that obtained in the case

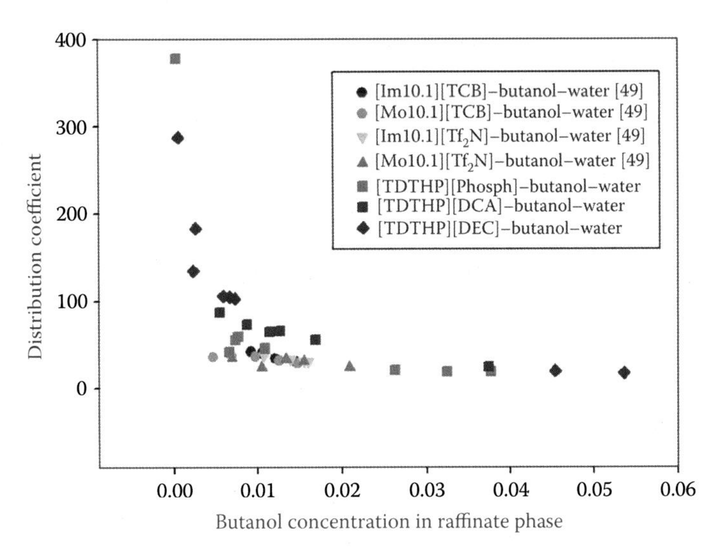

FIGURE 2.10
Comparison of the distribution coefficient with imidazolium-based cations at ambient conditions.

of 1-hexyl-3-methylimidazolium bis (trifluoromethylsulfonyl) imide by Chapeaux et al. (2008). It is an indication of better separation with wide range of feed concentration. In addition, the solvent requirement or the solvent–feed ratio will be less as compared to imidazolium-based ILs for a particular degree of separation.

For the ternary system containing [TDTHP][Phosph], the mole fraction of butanol in the raffinate phase is very low resulting in higher selectivity (Figures 2.4 and 2.5). For the other two systems, the mole fraction of water in the extract phase is nearly zero; hence, selectivity approaches infinity. The ternary diagram (Figures 2.6 through 2.9) also shows negligible concentration of both IL and water in the raffinate phase and the extract phase, respectively. It implies that [TDTHP][DCA] and [TDTHP][DEC] are completely immiscible with water in the ternary system. This confirms the findings of Cascon and Choudhary (2013) for the IL [TDTHP][DCA], where SILM-based supported ionic liquid membrane pervaporation of 1-butanol indicated high affinity for alcohol. This is mainly due to the strong hydrogen bonding between the dicyanamide anion and 1-butanol. Hydrogen bond energy (Garbuz, Skopenko, Khavryuchenko, & Gerasimchuk, 1989) between dicyanamide ion and butanol (60.8 kJ/mole) was found to be more as compared to that with water (53.9 kJ/mole). This is contrary to the measurements by Freire et al. (2008), where [TDTHP][DCA] and [TDTHP][DEC] were found to be more miscible with water. However, the hydrogen bond between the anions and butanol also alters the miscibility of the ILs with butanol and water. It can be seen from Figures 2.4 through 2.9 that the water content in [TDTHP][Phosph] is more as compared to that in [TDTHP][DCA] and [TDTHP][DEC]. This is attributed to the more electronegative atoms (one phosphorous and two oxygen atoms) in [Phosph] anion, thereby forming a strong bond with water (Garbuz et al. 1989).

We will now shift our attention to the second formulation system, namely 'Extraction of Biochemicals (Acetic Acid and Furfural) from Aqueous Phase' which is discussed in Section 2.1.2. One of the primary components namely acetic acid shall be first investigated. For acetic acid, the LLE experiments gave the distribution coefficient in the range 1.8–2.6. These values are much higher than 0.024–0.35 as obtained by Matsumoto et al. (2004) with imidazolium-based ILs such as [C_4mim][PF_6], [C_6mim][PF_6] and [C_8mim][PF_6]. The distribution coefficient for furfural was in the range 85–165. It suggests furfural can be extracted more efficiently from the aqueous solution in comparison to acetic acid. Further, the sloping of the tie lines for furfural (Figures 2.11 and 2.12) was found higher in magnitude and more positive as compared to acetic acid (Figures 2.13 and 2.14). This provides a pathway for selective simultaneous separation of furfural and acetic acid from aqueous phases. For acetic acid, selectivity is 4.2–7.4, while for furfural it varies from 316 to 544, which indicates a better ability of IL to extract furfural as compared to acetic acid from the aqueous phase. Pei et al. (2008) have also reported the extraction of furfural and acetic acid from aqueous solution using [C_4mim][PF_6], [C_6mim][PF_6] and [C_8mim][PF_6] and concluded that the

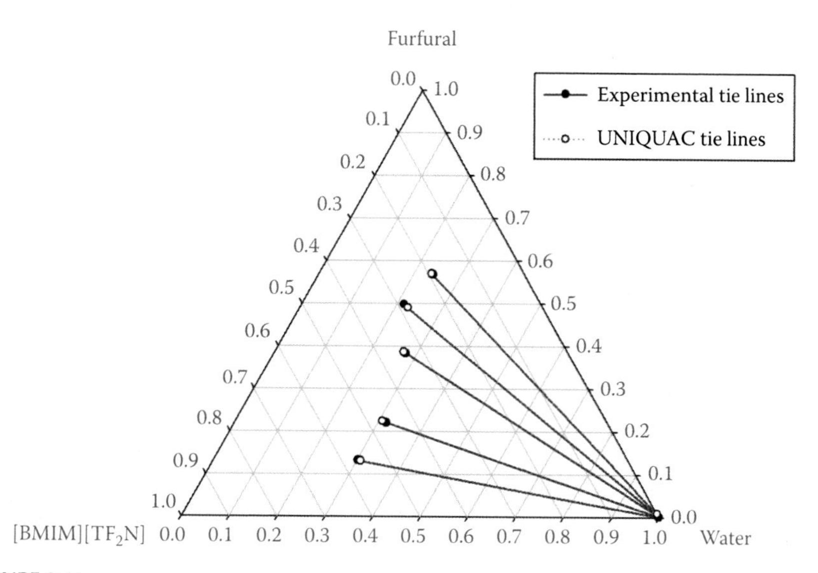

FIGURE 2.11
Experimental and UNIQUAC-predicted tie lines for the ternary system [BMIM][TF_2N]–furfural–water at $T = 298.15$ K and $p = 1$ atm.

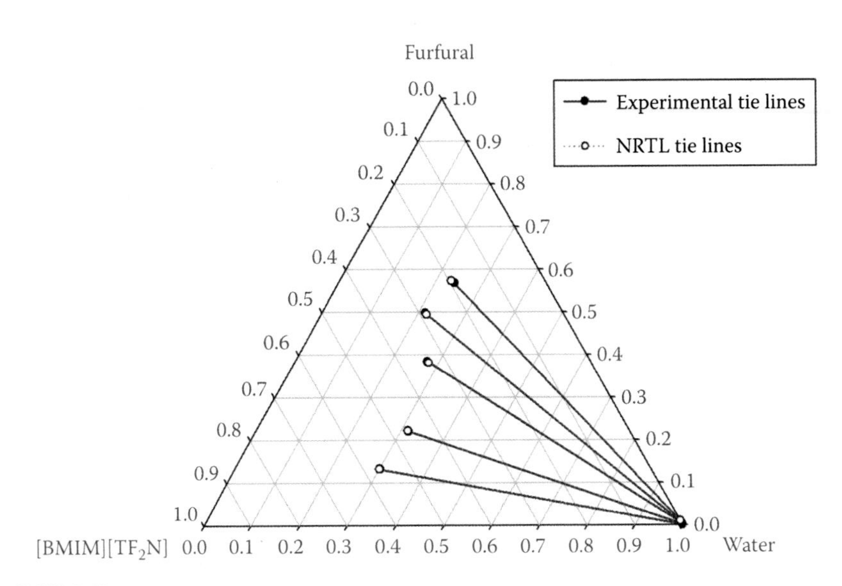

FIGURE 2.12
Experimental and NRTL-predicted tie lines for the ternary system [BMIM][TF$_2$N]–furfural–water at $T = 298.15$ K and $p = 1$ atm.

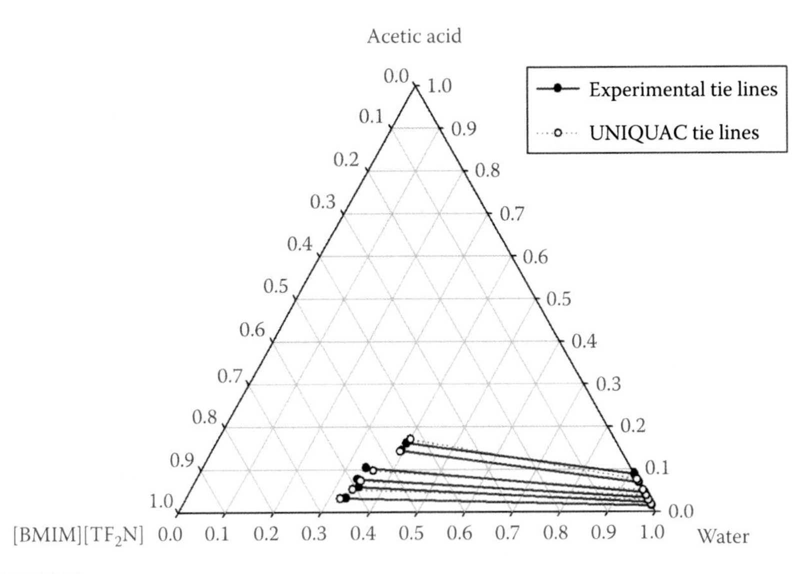

FIGURE 2.13
Experimental and UNIQUAC-predicted tie lines for the ternary system [BMIM][TF$_2$N]–acetic acid–water at $T = 298.15$ K and $p = 1$ atm.

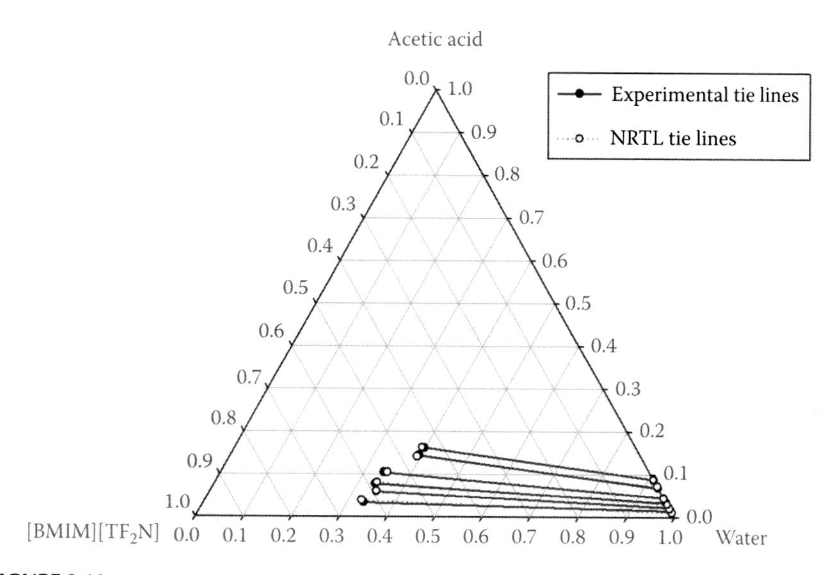

FIGURE 2.14
Experimental and NRTL-predicted tie lines for the ternary system [BMIM][TF$_2$N]–acetic acid–water at $T = 298.15$ K and $p = 1$ atm.

partition coefficients of furfural are higher significantly than those of acetic acid. The separation factor was found to be around 50. Pei et al. (2008) have also compared the partition coefficient of ILs with conventional solvents. The partition coefficients of furfural into the conventional organic solvents are usually much lower than those into the ILs (15.0–16.2). The analysis of the raffinate phase (aqueous phase) reveals negligible amount of IL in the aqueous phase. This makes the use of IL as a solvent economical as the loss of the solvent to the aqueous phase is negligible. Another important finding of this experimental work is that [BMIM][TF$_2$N] can be used to extract acetic acid for mole fractions of acetic acid less than 0.15 in the aqueous solution. This is usually the limit of acetic acid in a typical bio-oil-derived aqueous phase (Vitasari et al., 2011). Similarly, furfural can be extracted from an aqueous solution for mole fractions of furfural lesser than 0.5 which is again the typical limit found in bio-oil-derived aqueous phase (Vitasari et al., 2011).

The calculated interaction parameter values, objective function and RMSD values for all the systems are given in Table 2.2. Tie lines generated by the NRTL and the UNIQUAC models were approximately similar to the experimental tie lines. Using the NRTL model, the RMSD values were found to be 0.28%, 0.12% and 0.14% for systems containing [TDTHP][Phosph], [TDTHP][DCA] and [TDTHP][DEC] respectively with UNIQUAC model. Similarly, the RMSD values were 0.55%, 0.48% and 0.55% for [TDTHP][Phosph], [TDTHP][DCA] and [TDTHP][DEC] respectively with UNIQUAC model. While using the NRTL model, RMSD values were

TABLE 2.2

NRTL/UNIQUAC Interaction Parameters for Ternary Systems at $T = 298.15$ K and $p = 1$ atm

	NRTL Model Parameters				UNIQUAC Model Parameters			
i–j	τ_{ij}	τ_{ji}	**Obj**	**%RMSD**	A_{ij}/K	A_{ji}/K	**Obj**	**%RMSD**
[TDTHP][Phosph] (1)–1-Butanol (2)–Water (3)								
1–2	15.888	−9.037	-3.8×10^{-4}	0.28	−76.898	55.056	-1.48×10^{-3}	0.55
1–3	14.496	8.471			569.69	−166.37		
2–3	16.906	3.219			395.36	27.333		
[TDTHP][DCA] (1)–1-Butanol (2)–Water (3)								
1–2	−1.93	10.45	-7.24×10^{-5}	0.12	−0.21	−155.56	-1.12×10^{-3}	0.48
1–3	15.71	10.47			1000	−51.98		
2–3	20	4.56			1000	−15.75		
[TDTHP][DEC] (1)–1-Butanol (2)–Water (3)								
1–2	−4.77	17.99	-8.75×10^{-5}	0.14	−232.11	38.2	-1.46×10^{-3}	0.55
1–3	17.24	5.45			1000	−87.58		
2–3	20	3.28			1000	−58.53		
[BMIM][TF$_2$N] (1)–Acetic acid (2) + Water (3)								
1–2	−1.8201	16.52	-4.5×10^{-4}	0.35	−266.06	−335.82	−0.002	0.75
1–3	10.582	14.226			248.66	127.29		
2–3	39.995	33.466			−273.86	−682.32		
[BMIM][TF$_2$N] (1)–Furfural (2) + Water (3)								
1–2	1.6956	−2.7893	-2×10^{-4}	0.26	−357.32	928.03	-5×10^{-4}	0.43
1–3	14.356	11.332			510.23	−86.744		
2–3	16.523	4.1444			631	−52.788		

0.35% and 0.26% for systems containing acetic acid and furfural, respectively. Similarly, the RMSD values were 0.75% and 0.43% for acetic acid and furfural, respectively, with the UNIQUAC model. Both models gave excellent fit as all the tie lines closely overlap with the experiment tie lines.

2.6.2 Quaternary Systems for Butanol Recovery

As mentioned in the previous section concerning the separation of butanol from an ABE fermenter, it is worth mentioning to investigate the effects of ethanol. Ethanol is present along with Butanol and hence needs to be quantified from LLE systems having four components, namely water, IL, ethanol and butanol. The butanol distribution coefficient (β_B) varied in the ranges 3.55–122.5 and 5.37–43.84 for [TDTHP][Phosph] and [TDTHP][DCA], respectively. The ethanol distribution coefficient (β_E) varied within 3.31–20.36 and 3.09–12.86 for [TDTHP][Phosph] and [TDTHP][DCA], respectively (Tables 2.3 and 2.4).

TABLE 2.3

Experimental Tie Lines and Feed Ratio of [TDTHP][Phosph] (1)–Ethanol (2)–1-Butanol (3)–Water (4) at $T = 298.15$ K and $p = 1$ atm

Sr.	IL-Rich Phase				Aqueous Phase									
No.	x_{IL}	$x_{butanol}$	$x_{ethanol}$	x_{water}	x_{IL}	$x_{butanol}$	$x_{ethanol}$	x_{water}	β_B	β_E	S_B	S_E	Mol Ratio (E/B)	Mol Ratio (W/IL)
1	0.109	0.469	0.359	0.063	0	0.012	0.065	0.923	39.08	5.52	572.60	80.92	1.569	36.324
2	0.042	0.497	0.423	0.038	0	0.022	0.056	0.922	22.59	7.55	548.13	183.27	0.785	36.324
3	0.221	0.245	0.398	0.137	0	0.002	0.039	0.958	122.50	10.21	856.61	71.36	2.615	19.373
4	0.051	0.649	0.273	0.027	0	0.016	0.042	0.943	40.56	6.50	1416.68	227.02	0.549	26.907
5	0.110	0.259	0.566	0.065	0	0.073	0.171	0.756	3.55	3.31	41.27	38.50	4.969	55.883
6	0.006	0.424	0.538	0.033	0	0.024	0.144	0.832	17.67	3.74	445.41	94.20	3.138	36.324
7	0.042	0.361	0.550	0.047	0	0.041	0.101	0.858	8.80	5.45	160.74	99.41	1.569	48.432
8	0.023	0.244	0.696	0.037	0	0.034	0.191	0.775	7.18	3.64	150.32	76.33	3.661	72.648
9	0.015	0.456	0.509	0.02	0	0.016	0.025	0.959	28.50	20.36	1366.58	976.26	0.271	36.324
10	0.004	0.516	0.473	0.007	0	0.019	0.025	0.956	27.16	18.92	3708.99	2583.93	0.268	72.648
11	0.009	0.672	0.296	0.023	0	0.014	0.021	0.965	48.00	14.10	2013.91	591.39	0.268	60.540

TABLE 2.4

Experimental Tie Lines and Feed Ratio of [TDTHP][DCA] (1)–Ethanol (2)–1-Butanol (3)–Water (4) at $T = 298.15$ K and $p = 1$ atm

Sr.	IL-Rich Phase				Aqueous Phase								Mol Ratio	Mol Ratio
No.	x_{IL}	$x_{butanol}$	$x_{ethanol}$	x_{water}	x_{IL}	$x_{butanol}$	$x_{ethanol}$	x_{water}	β_B	β_E	S_B	S_E	(E/B)	(W/IL)
1	0.157	0.464	0.379	0	0	0.027	0.075	0.898	17.19	5.05	–	–	1.569	25.458
2	0.107	0.614	0.28	0	0	0.025	0.061	0.914	24.56	4.59	–	–	0.785	25.458
3	0.367	0.267	0.366	0	0	0.007	0.057	0.935	38.14	6.42	–	–	2.615	13.578
4	0.096	0.653	0.251	0	0	0.023	0.05	0.927	28.39	5.02	–	–	0.549	18.858
5	0.068	0.773	0.159	0	0.008	0.144	0.046	0.802	5.37	3.46	–	–	0.330	23.761
6	0.142	0.289	0.568	0	0.001	0.044	0.159	0.797	6.57	3.57	–	–	3.138	25.458
7	0.103	0.458	0.439	0	0	0.051	0.126	0.824	8.98	3.48	–	–	1.569	33.944
8	0.151	0.276	0.572	0	0	0.029	0.185	0.786	9.52	3.09	–	–	4.969	39.167
9	0.042	0.833	0.125	0	0	0.019	0.021	0.96	43.84	5.95	–	–	0.268	36.369
10	0.053	0.793	0.154	0	0	0.02	0.034	0.946	39.65	4.53	–	–	0.268	72.738
11	0.018	0.712	0.27	0	0	0.018	0.021	0.961	39.56	12.86	–	–	0.268	60.615

For [TDTHP][Phosph] containing system, the butanol selectivity varied within 41.27–3708.99 which is higher than ethanol selectivity, that is, 38.50–2583.93.

In the case of a system containing [TDTHP][DCA], the water in the extract phase was negligible, that is, within uncertainty values, thereby resulting in infinite selectivity. Hence, the results are not included in Table 2.4. For a system containing [TDTHP][Phosph], the largest β_B (122.50) was found as 2.615 for ethanol/butanol molar ratio and 19.373 for water/[TDTHP][Phosph] molar ratio. Similarly, the highest value of β_E was obtained at 0.268 for ethanol/butanol molar ratio and 36.324 for water/[TDTHP][Phosph] molar ratio. Water/[TDTHP][Phosph] molar ratio of 72.648 gave a higher S_B (3708.99) and S_E (2583.93) for the ethanol-to-butanol molar ratio of 0.268. There was not much improvement in β_B (39.56–43.84) with varying water/[TDTHP][DCA] molar ratio at a constant ethanol/butanol molar ratio of 0.268. On similar lines, β_E gave the largest value (12.86) for ethanol/butanol molar ratio of 0.268 and water/[TDTHP][DCA] molar ratio of 60.615. It can be observed that β_B is always higher than β_E for both systems. It indicates both ILs are more selective for butanol separation than ethanol. This is primarily due to the lesser polarity index of butanol (4) than ethanol (5.2), hence dissolving easily in non-polar ILs. β_B was found to be 12 times higher than β_E for the lowest mole ratio (19.4) of water/[TDTHP][Phosph]. For the lowest mole ratio (13.6) of water/[TDTHP][DCA], β_B was six times higher than β_E. In both cases, ethanol-to-butanol feed mole ratio was maintained at 2.6. For a constant value of water/[TDTHP][Phosph] molar ratio of 36.32, β_E was found to increase with decreasing ethanol-to-butanol feed ratio. However, β_B did not follow any specific trend. It was noted that β_B was seven times higher than β_E at ethanol-to-butanol feed molar ratio of 1.57. Similarly, for water/[TDTHP][DCA] molar ratio of 25.46, β_B increased with decreasing ethanol-to-butanol feed molar ratio. β_B was also 5.4 times higher than β_E at ethanol to butanol molar ratio of 0.78. These values are much higher than those obtained by Kubiczek and Kamiński (2013) for 1-hexyl-3-methylimidazolium hexafluorophosphate and 1-butyl-3-methylimidazolium bis (trifluoromethylsulfonyl) imide ILs.

The calculated interaction parameter values, objective function and RMSD values for all the systems are also given in Table 2.5. Tie lines generated by the NRTL and the UNIQUAC models were approximately similar to the experimental tie lines. Using the NRTL model, the RMSD values were 0.47% and 0.23% for systems containing [TDTHP][Phosph] and [TDTHP][DCA] (Figures 2.15 and 2.16). The corresponding RMSD values were 0.80% and 0.76% for [TDTHP][Phosph] and [TDTHP][DCA] respectively with UNIQUAC model.

TABLE 2.5

NRTL and UNIQUAC Interaction Parameters for Quaternary Systems at $T = 298.15$ K and $p = 1$ atm

	NRTL Model Parameters				UNIQUAC Model Parameters			
i–j	τ_{ij}	τ_{ji}	Obj	%RMSD	A_{ij}/K	A_{ji}/K	Obj	%RMSD
[TDTHP][Phosph] (1)–Ethanol (2)–1-Butanol (3)–Water (4)								
1–2	19.85	17.20	-1.42×10^{-3}	0.47	812.47	−621.74	-4.12×10^{-3}	0.80
1–3	19.98	1.76			769.88	−603.32		
1–4	8.41	6.83			1493.6	−610.46		
2–3	6.97	2.07			−83.66	960.18		
2–4	14.15	5.07			894.60	−17.90		
3–4	11.11	3.47			754.54	77.52		
[TDTHP][DCA] (1)–Ethanol (2)–1-Butanol (3)–Water (4)								
1–2	27.02	−27.60	-3.5×10^{-4}	0.23	−276.38	252.26	-3.68×10^{-3}	0.76
1–3	4.14	−12.17			−613	409.74		
1–4	9.97	−17.40			1495.30	−6.67		
2–3	1.21	3.22			−570.56	732.36		
2–4	11.33	3.49			1249.40	−86.93		
3–4	10.36	3.76			1114.40	−448.59		

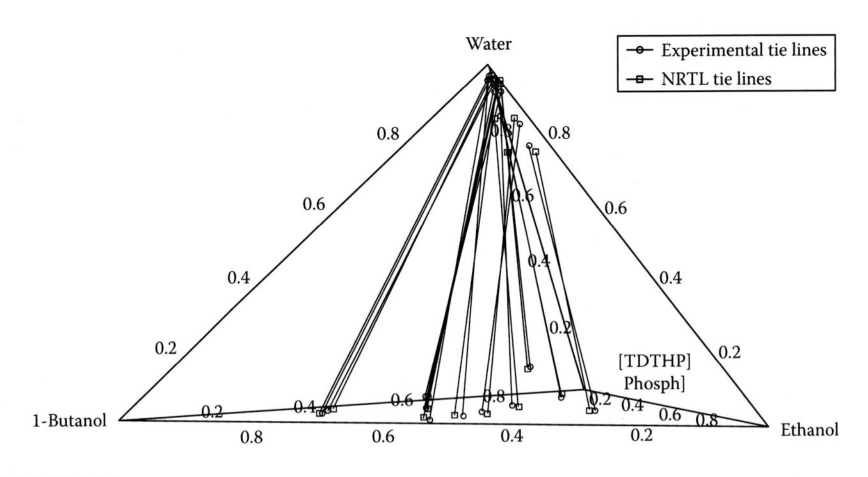

FIGURE 2.15
Experimental tie lines for the system [TDTHP][Phosph] (1)–Ethanol (2)–1-Butanol (3)–Water (4) at $T = 298.15$ K and $p = 1$ atm.

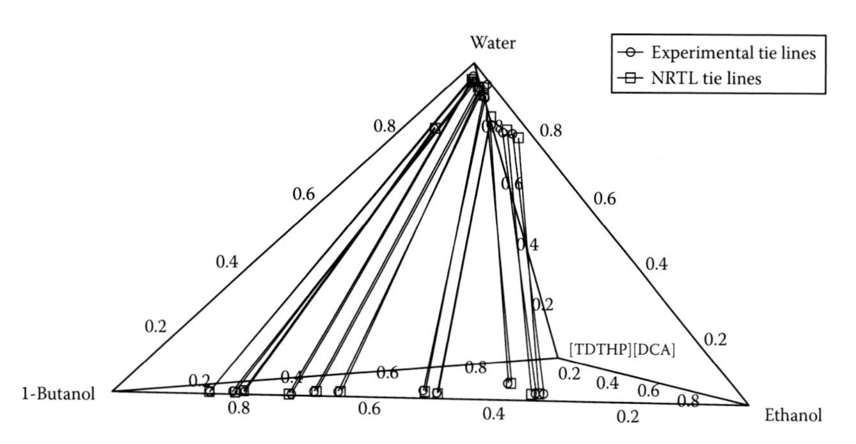

FIGURE 2.16
Experimental tie lines for the system [TDTHP][DCA] (1)–Ethanol (2)–1-Butanol (3)–Water (4) at $T = 298.15$ K and $p = 1$ atm.

2.7 Prediction of Phase Behavior by Statistical Associating Fluid Theory Models

EoS play an important role in chemical and biochemical engineering process design. They have assumed an important place in the prediction of phase equilibria of fluids and fluid mixtures. Originally, EoS were used mainly for pure components. Later on it was applied to mixtures of nonpolar components and subsequently extended for the prediction of phase equilibria of polar mixtures. The history of EoS goes back to the year 1873 with the development of van der Waals EoS. Based on van der Waals' EoS, researchers have developed various EoS such as Redlich–Kwong, Soave–Redlich–Kwong, Peng–Robinson, Guggenheim and Carnahan–Starling (Wei & Sadus, 2000) to name a few. Advances in statistical mechanics allowed the development of EoS based on molecular principles that are able to describe thermodynamic properties of real fluids accurately. Using Wertheim's first-order thermodynamic perturbation theory (TPT-1; Wertheim, 1984a, 1984b, 1986a, 1986b, 1986c, 1987), Chapman, Gubbins, Jackson and Radosz (1989, 1990) and Huang and Radosz (1990, 1991) developed the SAFT which is an accurate EoS for pure fluids and mixtures containing associating fluids.

In the SAFT model, molecules are considered to be composed of equal-size, spherical segments. Numbers of segments and segment diameters vary with the molecule size and shape. For a pure component, the fluid is first assumed to consist of equal-sized hard spheres. Next, the segment–segment attractive forces are added to each sphere. Thereafter, the chain sites are added to each sphere and chain molecules are formed by the bonding of chain sites. Finally, specific association sites are added at the same position through attractive interactions and contribution to Helmholtz free energy for each step is evaluated. Therefore, in the SAFT model, the residual Helmholtz energy per molecule for a pure component has a hard sphere ($\tilde{a}^{hs}$), dispersion ($\tilde{a}^{disp}$), chain ($\tilde{a}^{chain}$) and an association ($\tilde{a}^{assoc}$) contribution and is given by

$$\tilde{a}^{res} = \tilde{a}^{hs} + \tilde{a}^{disp} + \tilde{a}^{chain} + \tilde{a}^{assoc} \tag{2.11}$$

where $\tilde{a}^{res}$ is the residual Helmholtz free energy of the system ($\tilde{a}^{res} = \tilde{a}^{total} - \tilde{a}^{ideal}$) and $\tilde{a} = A/NkT$.

A variant, namely the PC-SAFT EoS was developed in 2001 by Gross and Sadowski (2001, 2002). Using hard-chain reference fluid and applying a perturbation theory for chain molecules, Gross and Sadowski (2001, 2002) derived a dispersion expression for chain molecules. PC-SAFT uses the hard-chain fluid as the reference system, whereas in SAFT, a hard-sphere fluid is considered as a reference system. In this EoS, molecules are assumed to be chains composed of hard spheres, which repel each other. The attractive forces among molecules are accounted by adding perturbation terms to the reference system which includes dispersive forces and specific associative interactions (Figure 2.17). The residual Helmholtz energy can now be calculated as the Helmholtz energy of the reference

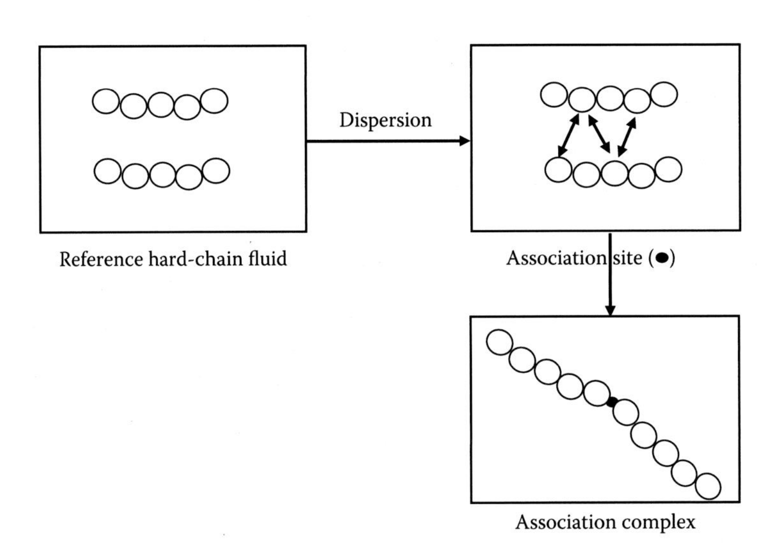

FIGURE 2.17
Procedure to form a molecule in the PC-SAFT model.

system ($\tilde{a}^{hc}$) superposed with the Helmholtz energy of perturbation, that is, dispersion ($\tilde{a}^{disp}$) and association ($\tilde{a}^{assoc}$):

$$\tilde{a}^{res} = \tilde{a}^{hc} + \tilde{a}^{disp} + \tilde{a}^{assoc} \quad (2.12)$$

Hard-Chain Contribution

The Helmholtz energy of the hard-chain reference term is given as

$$\tilde{a}^{hc} = \frac{A^{hc}}{NkT} = \bar{m}\frac{A^{hs}}{N_s kT} - \sum_i x_i(m_i - 1)\ln g_{ii}^{hs}(\mathrm{d}_{ii}) \quad (2.13)$$

where x_i is the mole fraction of chains of component i, m_i is the number of segments in a chain and the mean segment number in the mixture is defined as

$$\bar{m} = \sum_i x_i m_i \quad (2.14)$$

The Helmholtz energy for the hard-sphere segments is given on a per-segment basis as

$$\frac{A^{hs}}{N_s kT} = \frac{1}{\xi_o}\left[\frac{3\xi_1\xi_2}{(1-\xi_3)} + \frac{\xi_2^3}{\xi_3(1-\xi_3)^2} + \left(\frac{\xi_2^3}{\xi_3^2} - \xi_o\right)\ln(1-\xi_3)\right] \quad (2.15)$$

where N_s is related to the number of hard spheres.

The radial pair distribution function for the hard-sphere fluid is given by

$$g_{ij}^{hs}(d_{ij}) = \frac{1}{(1-\xi_3)} + \left(\frac{d_i d_j}{d_i + d_j}\right)\frac{3\xi_2}{(1-\xi_3)^2} + \left(\frac{d_i d_j}{d_i + d_j}\right)^2 \frac{2\xi_2^2}{(1-\xi_3)^3} \quad (2.16)$$

and ξ_n is defined as

$$\xi_n = \frac{\pi}{6}\rho\sum_i x_i m_i d_i^n \qquad n = \{0,1,2,3\} \quad (2.17)$$

The temperature-dependent segment diameter is obtained as

$$d_i(T) = \sigma_i\left[1 - 0.12\exp\left(\frac{-3\varepsilon_i}{kT}\right)\right] \quad (2.18)$$

where:

σ_i is the temperature-independent segment diameter

ε_i/k is the depth of the pair-potential

Dispersion Contribution

The dispersion contribution to the Helmholtz energy is given by

$$\frac{A^{\text{disp}}}{NkT} = -2\pi\rho I_1(\eta,\bar{m})\sum_i\sum_j x_i x_j m_i m_j\left(\frac{\varepsilon_{ij}}{kT}\right)\sigma_{ij}^3 \\ -\pi\rho\bar{m}C_1 I_2(\eta,\bar{m})\sum_i\sum_j x_i x_j m_i m_j\left(\frac{\varepsilon_{ij}}{kT}\right)^2\sigma_{ij}^3 \tag{2.19}$$

with

$$C_1 = \left[1 + Z^{\text{hc}} + \rho\frac{\partial Z^{\text{hc}}}{\partial\rho}\right]^{-1} \\ = \left[1 + \bar{m}\left(\frac{8\eta - 2\eta^2}{(1-\eta)^4}\right) + (1-\bar{m})\left(\frac{20\eta - 27\eta^2 + 12\eta^3 - 2\eta^4}{[(1-\eta)(2-\eta)]^2}\right)\right]^{-1} \tag{2.20}$$

Power series I_1 and I_2 depend only on density and segment number according to the following equations:

$$I_1(\eta,\bar{m}) = \sum_{i=0}^{6} a_i(\bar{m})\eta^i \tag{2.21}$$

$$I_2(\eta,\bar{m}) = \sum_{i=0}^{6} b_i(\bar{m})\eta^i \tag{2.22}$$

where the coefficients a_i(m) and b_i(m) are functions of the segment number:

$$a_i(\bar{m}) = a_{oi} + \frac{\bar{m}-1}{\bar{m}}a_{1i} + \frac{\bar{m}-1}{\bar{m}}\frac{\bar{m}-2}{\bar{m}}a_{2i} \tag{2.23}$$

$$b_i(\bar{m}) = b_{oi} + \frac{\bar{m}-1}{\bar{m}}b_{1i} + \frac{\bar{m}-1}{\bar{m}}\frac{\bar{m}-2}{\bar{m}}b_{2i} \tag{2.24}$$

Association Contribution

The association contribution to the Helmholtz energy is given as

$$\frac{A^{\text{assoc}}}{NkT} = \sum_i x_i \sum_{A_i=1}^{n\text{site}}\left(\ln X^{A_i} - \frac{X^{A_i}}{2} + \frac{M_i}{2}\right) \tag{2.25}$$

where X^{A_i} is the fraction of the free molecules i that are not bonded at the association site A:

$$X^{A_i} = \left(1+\rho\sum_j x_j \sum_{B_j}^{nsites} X^{B_j}\Delta^{A_iB_j}\right)^{-1} \tag{2.26}$$

with

$$\Delta^{A_iB_j} = g_{ij}^{hs}(d_{ij})\bullet \kappa^{A_iB_j}\bullet d_{ij}^3\left(\exp\left(\frac{\varepsilon^{A_iB_j}}{kT}\right)-1\right) \tag{2.27}$$

In terms of the compressibility factor Z, the EoS is given as the sum of the ideal gas contribution ($Z^{id} = 1$), the hard-chain contribution (Z^{hc}), the dispersion (attractive) contribution (Z^{disp}) and the contribution due to associating interactions (Z^{assoc}). Thus,

$$Z = Z^{id} + Z^{hc} + Z^{disp} + Z^{assoc} \tag{2.28}$$

The expressions for all the terms are given in Appendix A. Pressure can be calculated by applying the following relation:

$$P = ZkT\rho \tag{2.29}$$

2.7.1 LLE Phase Equilibria Computation

LLE calculations of a multi-component system are based on equal fugacity values, f_i, of all components, i, in phases I and II and are performed using the relation

$$x_i^I\varphi_i^I = x_i^{II}\varphi_i^{II} \tag{2.30}$$

where x_i and φ_i are the mole fraction and fugacity coefficient, respectively, of component i. The expression for calculating the fugacity coefficient for component i is given in Appendix A.

PC-SAFT parameters of pure IL 1-butyl-3-methylimidazolium bis (trifluoromethylsulfonyl) imide ([BMIM][Tf_2N]) were first determined by considering the IL as a nonassociating compound. The three pure-component parameters m, σ and ε/k_B were obtained by fitting to experimental density data from the literature (Tariq et al., 2010). In the second step, the IL was considered as a self-associating molecule. Two additional pure-component parameters determine the associating interactions between association sites

A_i and B_j of a pure component i. These are the association energy ε^{AiBj}/k and the effective association volume κ^{AiBj}. The associating parameters, ε^{AiBj}/k = 3450 K and κ^{AiBj} = 0.00225, were kept constant to reduce the number of parameters to be fitted (Andreu & Vega, 2008). The IL was then modelled as a molecule with two associating sites (2B association scheme). This implies that the values of $\Delta^{AA} = \Delta^{BB} = 0$, $\Delta^{AB} \neq 0$ and $X^A = X^B$ with the non-bonding fraction as (Huang & Radosz, 1990)

$$X^A = \frac{-1 + (1 + 4\rho\Delta)^{1/2}}{2\rho\Delta} \tag{2.31}$$

The following objective function was used in both strategies to determine the pure IL parameters:

$$\mathrm{OF} = \sum_{i=1}^{n_{\mathrm{pts}}} \left(1 - \frac{\rho_i^{\mathrm{calc}}}{\rho_i^{\mathrm{exp}}}\right)^2 \tag{2.32}$$

This takes into account the deviations between calculated (calc) and experimental (exp) liquid densities. The percentage absolute average relative deviation between the experimentally obtained and calculated densities was determined according to (average absolute relative deviation [AARD], %):

$$\mathrm{AARD}(\%) = \frac{100}{n_{\mathrm{pts}}} \sum_{i=1}^{n_{\mathrm{pts}}} \left|1 - \frac{\rho_i^{\mathrm{calc}}}{\rho_i^{\mathrm{exp}}}\right| \tag{2.33}$$

The temperature–density plot for $[BMIM][TF_2N]$ considered as nonassociating and self-associating is shown in Figure 2.18. As seen in Figure 2.18, the agreement between the experimental data and the prediction with PC-SAFT is excellent. However, deviation at a higher temperature is more. This may be due to the fact that the considered IL is nonassociating as compared to a self-associating compound. PC-SAFT parameters obtained after optimization are provided in Table 2.6. The AARD (%) for IL is 0.22% as a nonassociating molecule, whereas as self-associating, AARD (%) is 0.12%. Thus, there is improvement in prediction when considering IL as self-associating.

For modelling mixture properties, the segment diameter and the dispersion energy parameter were estimated by applying the combining rules of Lorentz and Berthelot:

$$\sigma_{ij} = \frac{1}{2}(\sigma_i + \sigma_j) \tag{2.34}$$

$$\varepsilon_{ij} = \sqrt{\varepsilon_i \varepsilon_j}(1 - k_{ij}) \tag{2.35}$$

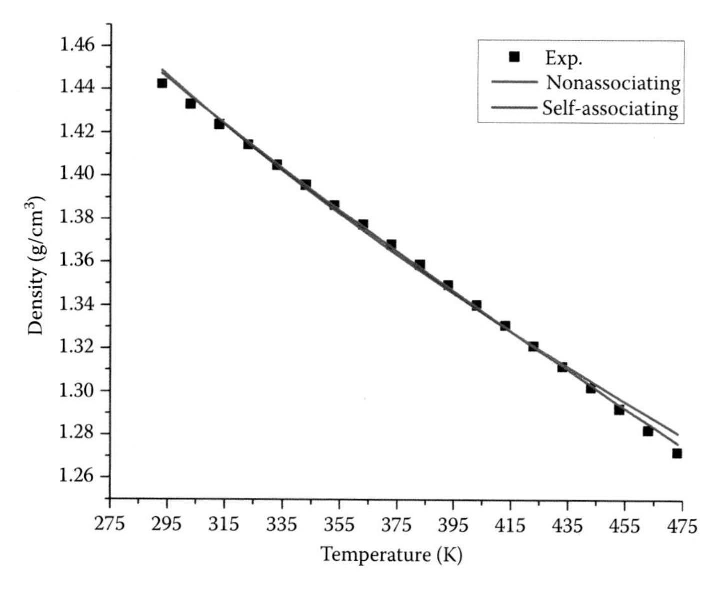

FIGURE 2.18
Temperature-density plot for [BMIM][TF_2N] considered as nonassociating and self-associating mode.

TABLE 2.6

Optimized PC-SAFT Parameters of Pure Components

Component	Scheme	m	σ (Å)	ε/k_B (K)	ε^{AiBj}/k (K)	κ^{AiBj}	AARD (%)
[BMIM][Tf_2N]	Nonassociating	7.27	3.95	382.00	–	–	0.2190
	Self-associating	5.28	4.41	383.68	3450.0	0.00225	0.1221
Acetic acid	Self-associating	1.32	3.84	201.48	3000.0	0.05	0.15
Water	Self-associating	1.02	2.97	399.39	2022.4	0.03	0.05
Furfural	Self-associating	2.89	3.42	256.94	2114.1	0.04	0.01

where k_{ij} is binary interaction parameters. To make the model predictive, k_{ij} was considered to be zero.

The cross-association parameters were calculated as suggested by Wolbach and Sandler (1998)

$$\varepsilon^{A_iB_j} = \frac{1}{2}(\varepsilon^{A_iB_i} + \varepsilon^{A_jB_j}) \tag{2.36}$$

$$k^{A_iB_j} = \sqrt{k^{A_iB_i}k^{A_jB_j}}\left[\frac{\sqrt{\sigma_i\sigma_j}}{\frac{1}{2}(\sigma_i+\sigma_j)}\right]^3 \tag{2.37}$$

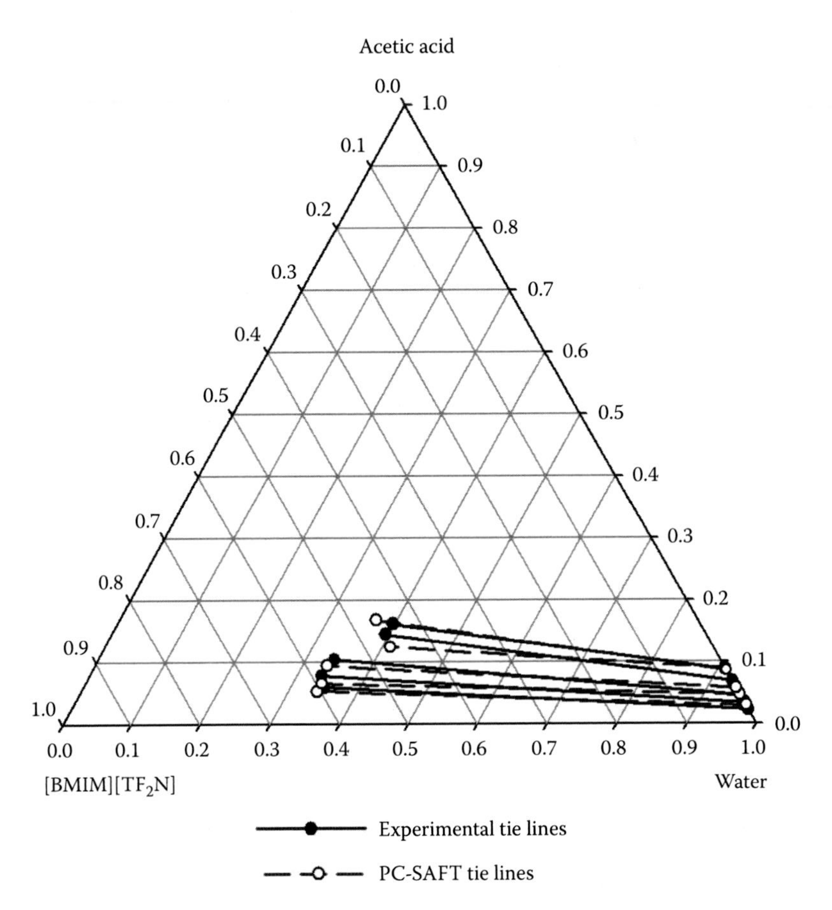

FIGURE 2.19
Experimental and PC-SAFT predicted tie lines for the ternary system [BMIM][TF_2N]–acetic acid–water at $T = 298.15$ K and $p = 1$ atm.

The LLE behavior of the ternary systems, [BMIM][Tf_2N] + Acetic acid + water and [BMIM][Tf_2N] + furfural + water, predicted by PC-SAFT EoS, are shown in Figures 2.19 and 2.20, respectively. The predicted behavior was in good agreement with the experimental data having root-mean-square deviation (RMSD) 1.19% for an acetic acid-based system, whereas 2.72% for a furfural-based system. This confirms the ability of PC-SAFT to correctly predict the phase behavior of ternary systems.

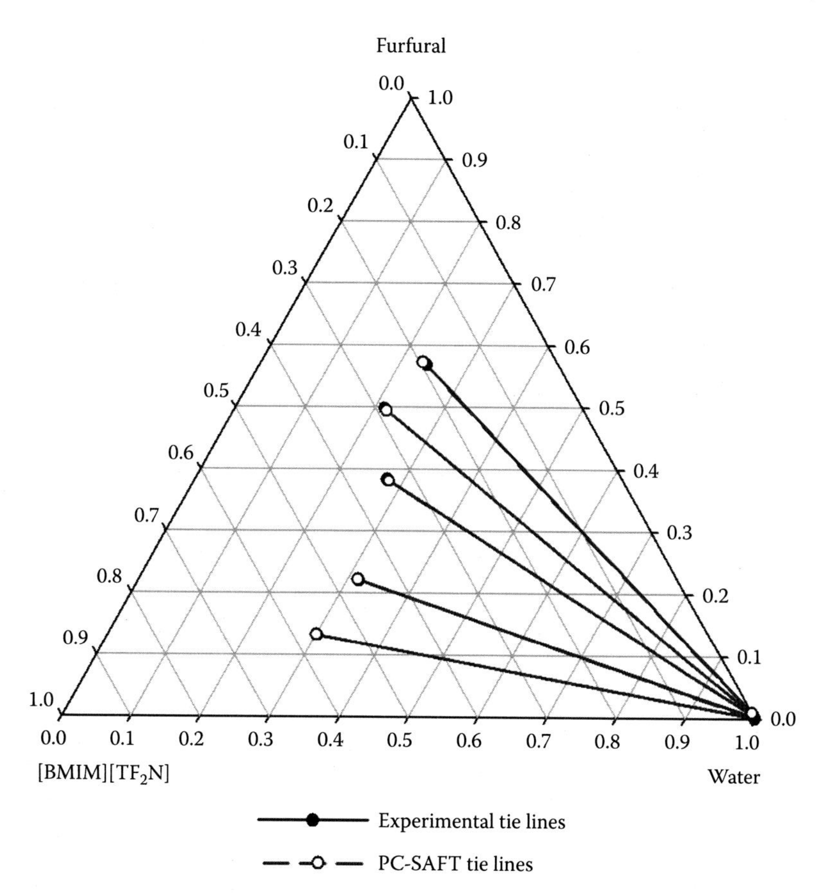

FIGURE 2.20
Experimental and PC-SAFT predicted tie lines for the ternary system [BMIM][TF_2N]–furfural–water at $T = 298.15$ K and $p = 1$ atm.

Appendix A

The expression for different contributing terms for the compressibility factor is as follows:

Hard-Chain Contribution

Hard-chain contribution to compressibility factor is given by

$$Z^{hc} = \bar{m}Z^{hs} - \sum_i x_i(m_i - 1)(g_{ii}^{hs})^{-1}\rho\frac{\partial g_{ii}^{hs}}{\partial \rho} \tag{A.1}$$

where:

x_i is the mole fraction of chains of component i
m_i is the number of segments in a chain of component i
ρ is the total number of density of molecules

The mean segment number ($\bar{m}$) in the mixture is defined as

$$\bar{m} = \sum_i x_i m_i \tag{A.2}$$

Z^{hs} is the contribution of the hard-sphere fluid given by

$$Z^{hs} = \frac{\xi_3}{1-\xi_3} + \frac{3\xi_1\xi_2}{\xi_0(1-\xi_3)^2} + \frac{3\xi_2{}^3 - \xi_3\xi_2{}^3}{\xi_0(1-\xi_3)^3} \tag{A.3}$$

The radial pair distribution function for the hard-sphere fluid is given by

$$g_{ij}^{hs}(d_{ij}) = \frac{1}{(1-\xi_3)} + \left(\frac{d_i d_j}{d_i + d_j}\right)\frac{3\xi_2}{(1-\xi_3)^2} + \left(\frac{d_i d_j}{d_i + d_j}\right)^2 \frac{2\xi_2^2}{(1-\xi_3)^3} \tag{A.4}$$

and ξ_n is defined as

$$\xi_n = \frac{\pi}{6}\rho \sum_i x_i m_i d_i^n \qquad n = \{0,1,2,3\} \tag{A.5}$$

The temperature-dependent segment diameter is obtained as

$$d_i(T) = \sigma_i\left[1 - 0.12\exp\left(\frac{-3\varepsilon_i}{kT}\right)\right] \tag{A.6}$$

where:

σ_i is the temperature-independent segment diameter
ε_i/k is the depth of the pair potential

$$\rho\frac{\partial g_{ij}^{hs}}{\partial \rho} = \frac{\xi_3}{(1-\xi_3)^2} + \left(\frac{d_i d_j}{d_i + d_j}\right)\left(\frac{3\xi_2}{(1-\xi_3)^2} + \frac{6\xi_2\xi_3}{(1-\xi_3)^3}\right) + \left(\frac{d_i d_j}{d_i + d_j}\right)^2\left(\frac{4\xi_2^2}{(1-\xi_3)^3} + \frac{6\xi_2^2\xi_3}{(1-\xi_3)^4}\right) \tag{A.7}$$

Dispersion Contribution

The dispersion contribution to compressibility factor is given by

$$Z^{\text{disp}} = -2\pi\rho \frac{\partial(\eta I_1)}{\partial\eta}\overline{m^2\varepsilon\sigma^3} - \pi\rho\bar{m}\left[C_1\frac{\partial(\eta I_2)}{\partial\eta} + C_2\eta I_2\right]\overline{m^2\varepsilon^2\sigma^3} \tag{A.8}$$

where

$$\frac{\partial(\eta I_1)}{\partial\eta} = \sum_{i=0}^{6} a_i(\bar{m})(i+1)\eta^i \tag{A.9}$$

$$\frac{\partial(\eta I_2)}{\partial\eta} = \sum_{i=0}^{6} b_i(\bar{m})(i+1)\eta^i \tag{A.10}$$

$$C_1 = \left[1 + \bar{m}\left(\frac{8\eta - 2\eta^2}{(1-\eta)^4}\right) + (1-\bar{m})\left(\frac{20\eta - 27\eta^2 + 12\eta^3 - 2\eta^4}{[(1-\eta)(2-\eta)]^2}\right)\right]^{-1} \tag{A.11}$$

$$C_2 = -C_1^2\left[\bar{m}\left(\frac{-4\eta^2 + 20\eta + 8}{(1-\eta)^5}\right) + (1-\bar{m})\left(\frac{2\eta^3 + 12\eta^2 - 48\eta + 40}{[(1-\eta)(2-\eta)]^3}\right)\right] \tag{A.12}$$

$$\overline{m^2\varepsilon\sigma^3} = \sum_i\sum_j x_i x_j m_i m_j\left(\frac{\varepsilon_{ij}}{kT}\right)\sigma_{ij}^3 \tag{A.13}$$

$$\overline{m^2\varepsilon^2\sigma^3} = \sum_i\sum_j x_i x_j m_i m_j\left(\frac{\varepsilon_{ij}}{kT}\right)^2\sigma_{ij}^3 \tag{A.14}$$

Power series I_1 and I_2 depend only on density and segment number according to:

$$I_1(\eta,\bar{m}) = \sum_{i=0}^{6} a_i(\bar{m})\eta^i \tag{A.15}$$

$$I_2(\eta,\bar{m}) = \sum_{i=0}^{6} b_i(\bar{m})\eta^i \tag{A.16}$$

where the coefficients $a_i(m)$ and $b_i(m)$ are functions of the segment number according to

$$a_i(\bar{m}) = a_{oi} + \frac{\bar{m}-1}{\bar{m}} a_{1i} + \frac{\bar{m}-1}{\bar{m}} \frac{\bar{m}-2}{\bar{m}} a_{2i} \quad (A.17)$$

$$b_i(\bar{m}) = b_{oi} + \frac{\bar{m}-1}{\bar{m}} b_{1i} + \frac{\bar{m}-1}{\bar{m}} \frac{\bar{m}-2}{\bar{m}} b_{2i} \quad (A.18)$$

and a_{oi}, a_{1i}, a_{2i}, b_{oi}, b_{1i} and b_{2i} are universal model constants.

Conventional combining rules are employed to determine the parameters for a pair of unlike segments:

$$\sigma_{ij} = \frac{(\sigma_i + \sigma_j)}{2} \quad (A.19)$$

$$\varepsilon_{ij} = \sqrt{\varepsilon_i \varepsilon_j}(1 - k_{ij}) \quad (A.20)$$

Association Contribution

The association contribution to compressibility factor is given by

$$Z^{\text{assoc}} = \sum_i x_i \rho \sum_A \left[\frac{1}{X^{Ai}} - \frac{1}{2} \right] \frac{\partial X^{A_i}}{\partial \rho} \quad (A.21)$$

where X^{Ai} is the fraction of molecules i that are not bonded at the association site A, which is written as

$$X^{A_i} = \left(1 + \rho \sum_j x_j \sum_{B_j}^{n\text{sites}} X^{Bj} \Delta^{A_i B_j} \right)^{-1} \quad (A.22)$$

with

$$\Delta^{A_i B_j} = g_{ij}^{\text{hs}}(d_{ij}) \cdot \kappa^{A_i B_j} \cdot d_{ij}^3 \left(\exp\left(\frac{\varepsilon^{A_i B_j}}{kT} \right) - 1 \right) \quad (A.23)$$

where Δ^{AiBj} is the association strength between two sites A and B belonging to two different molecules i and j.

Fugacity Coefficient

The fugacity coefficient is related to the residual chemical potential according to

$$\ln \phi_k = \frac{\mu_k^{\text{res}}(T,V)}{kT} - \ln Z \quad (A.24)$$

The chemical potential can be obtained from

$$\frac{\mu_k^{\text{res}}(T,V)}{kT}=\tilde{a}^{\text{res}}+(Z-1)+\left(\frac{\partial\tilde{a}^{\text{res}}}{\partial x_k}\right)_{T,V,x_{i\neq k}}-\sum_{j=1}^{N}\left[x_j\left(\frac{\partial\tilde{a}^{\text{res}}}{\partial x_j}\right)_{T,V,x_{i\neq j}}\right] \quad \text{(A.25)}$$

with

$$\xi_{n,xk}=\frac{\partial\xi_n}{\partial x_k}=\frac{\pi}{6}\rho m_k(d_k)^n \qquad n\in\{0,1,2,3\} \quad \text{(A.26)}$$

Hard-Chain Contribution

$$\left(\frac{\partial\tilde{a}^{\text{hc}}}{\partial x_k}\right)=m_k\tilde{a}^{\text{hs}}+\bar{m}\left(\frac{\partial\tilde{a}^{\text{hs}}}{\partial x_k}\right)_{T,\rho,x_{j\neq k}}-\sum_i x_i(m_i-1)(g_{ii}^{\text{hs}})^{-1}\left(\frac{\partial g_{ii}^{\text{hs}}}{\partial x_k}\right)_{T,\rho,x_{j\neq k}} \quad \text{(A.27)}$$

$$\left(\frac{\partial\tilde{a}^{\text{hs}}}{\partial x_k}\right)=-\frac{\xi_{o,x_k}}{\xi_o}\tilde{a}^{\text{hs}}+\frac{1}{\xi_0}\left[\begin{array}{l}\dfrac{3(\xi_{1,x_k}\xi_2+\xi_1\xi_{2,x_k})}{(1-\xi_3)}+\dfrac{3\xi_1\xi_2\xi_{3,x_k}}{(1-\xi_3)^2}+\dfrac{3\xi_2^2\xi_{2,x_k}}{\xi_3(1-\xi_3)^2}+\\ \dfrac{\xi_2^3\xi_{3,x_k}(3\xi_3-1)}{\xi_3^2(1-\xi_3)^3}+\\ \left(\dfrac{3\xi_2^2\xi_{2,x_k}\xi_3-2\xi_2^3\xi_{3,x_k}}{\xi_3^3}-\xi_{0,x_k}\right)\ln(1-\xi_3)\\ +\left(\xi_0-\dfrac{\xi_2^3}{\xi_3^2}\right)\dfrac{\xi_{3,x_k}}{(1-\xi_3)}\end{array}\right] \quad \text{(A.28)}$$

$$\begin{aligned}\frac{\partial g_{ij}^{\text{hs}}}{\partial x_k}&=\frac{\xi_{3,x_k}}{(1-\xi_3)^2}+\left(\frac{d_id_j}{d_i+d_j}\right)\left(\frac{3\xi_{2,x_k}}{(1-\xi_3)^2}+\frac{6\xi_2\xi_{3,x_k}}{(1-\xi_3)^3}\right)+\\ &\left(\frac{d_id_j}{d_i+d_j}\right)^2\left(\frac{4\xi_2\xi_{2,x_k}}{(1-\xi_3)^3}+\frac{6\xi_2^2\xi_{3,x_k}}{(1-\xi_3)^4}\right)\end{aligned} \quad \text{(A.29)}$$

Dispersion Contribution

$$\begin{aligned}\frac{\partial\tilde{a}^{\text{disp}}}{\partial x_k}&=-2\pi\rho\left[I_{1,x_k}\overline{m^2\varepsilon\sigma^3}+I_1\left(\overline{m^2\varepsilon\sigma^3}\right)_{x_k}\right]\\ &-\pi\rho\left\{\begin{array}{l}\left[m_kC_1I_2+\bar{m}C_{1,x_k}I_2+\bar{m}C_1I_{2,x_k}\right]\overline{m^2\varepsilon^2\sigma^3}\\ +\bar{m}C_1I_2(\overline{m^2\varepsilon^2\sigma^3})_{x_k}\end{array}\right\}\end{aligned} \quad \text{(A.30)}$$

with

$$\left(\overline{m^2\varepsilon\sigma^3}\right)_{xk} = 2m_k \sum_j x_j m_j \left(\frac{\varepsilon_{kj}}{kT}\right) \sigma_{kj}^3 \tag{A.31}$$

$$\left(\overline{m^2\varepsilon^2\sigma^3}\right)_{xk} = 2m_k \sum_j x_j m_j \left(\frac{\varepsilon_{kj}}{kT}\right)^2 \sigma_{kj}^3 \tag{A.32}$$

$$C_{1,xk} = C_2\xi_{3,xk} - C_1^2\left[m_k\left(\frac{8\eta-2\eta^2}{(1-\eta)^4}\right) - m_k\left(\frac{20\eta-27\eta^2+12\eta^3-2\eta^4}{[(1-\eta)(2-\eta)]^2}\right)\right] \tag{A.33}$$

$$I_{1,xk} = \sum_{i=0}^{6}\left[a_i(\overline{m})i\xi_{3,xk}\eta^{i-1} + a_{i,xk}\eta^i\right] \tag{A.34}$$

$$I_{2,xk} = \sum_{i=0}^{6}\left[b_i(\overline{m})i\xi_{3,xk}\eta^{i-1} + b_{i,xk}\eta^i\right] \tag{A.35}$$

$$a_{i,xk} = \frac{m_k}{\overline{m}^2}a_{1i} + \frac{m_k}{\overline{m}^2}\left(3 - \frac{4}{\overline{m}}\right)a_{2i} \tag{A.36}$$

$$b_{i,xk} = \frac{m_k}{\overline{m}^2}b_{1i} + \frac{m_k}{\overline{m}^2}\left(3 - \frac{4}{\overline{m}}\right)b_{2i} \tag{A.37}$$

References

Abrams, D. S., & Prausnitz, J. M. (1975). Statistical thermodynamics of liquid mixtures: A new expression for the excess Gibbs energy of partly or completely miscible systems. *AIChE Journal, 21*(1), 116–128. doi:10.1002/aic.690210115.

Andreu, J. S., & Vega, L. F. (2008). Modeling the solubility behavior of CO_2, H_2, and Xe in [Cn-mim][Tf_2N] ionic liquids. *The Journal of Physical Chemistry B, 112*(48), 15398–15406. doi:10.1021/jp807484g.

Arce, A., Earle, M. J., Katdare, S. P., Rodríguez, H., & Seddon, K. R. (2008). Application of mutually immiscible ionic liquids to the separation of aromatic and aliphatic hydrocarbons by liquid extraction: A preliminary approach. *Physical Chemistry Chemical Physics, 10*(18), 2538–2542. doi:10.1039/B718101A.

Arce, A., Earle, M. J., Katdare, S. P., Rodríguez, H., & Seddon, K. R. (2007). Phase equilibria of mixtures of mutually immiscible ionic liquids. *Fluid Phase Equilibria, 261*(1–2), 427–433. doi:10.1016/j.fluid.2007.06.017.

Arce, A., Earle, M. J., Rodriguez, H., Seddon, K. R., & Soto, A. (2009). Bis{(trifluoromethyl)sulfonyl}amide ionic liquids as solvents for the extraction of aromatic hydrocarbons from their mixtures with alkanes: Effect of the nature of the cation. *Green Chemistry, 11*(3), 365–372. doi:10.1039/B814189D.

Arce, A., Francisco, M., & Soto, A. (2010). Evaluation of the polysubstituted pyridinium ionic liquid [hmmpy][Ntf2] as a suitable solvent for desulfurization: Phase equilibria. *The Journal of Chemical Thermodynamics, 42*(6), 712–718. doi:10.1016/j.jct.2010.01.005.

Arce, A., Rodríguez, H., & Soto, A. (2006). Purification of ethyl tert-butyl ether from its mixtures with ethanol by using an ionic liquid. *Chemical Engineering Journal, 115*(3), 219–223. doi:10.1016/j.cej.2005.10.010.

Aznar, M. (2007). Correlation of (liquid + liquid) equilibrium of systems including ionic liquids. *Brazilian Journal of Chemical Engineering, 24*, 143–149.

Banerjee, T., Singh, M. K., Sahoo, R. K., & Khanna, A. (2005). Volume, surface and UNIQUAC interaction parameters for imidazolium based ionic liquids via polarizable continuum model. *Fluid Phase Equilibria, 234*(1–2), 64–76. doi:10.1016/j.fluid.2005.05.017.

Bennett, N. M., Helle, S. S., & Duff, S. J. B. (2009). Extraction and hydrolysis of levoglucosan from pyrolysis oil. *Bioresource Technology, 100*(23), 6059–6063. doi:10.1016/j.biortech.2009.06.067.

Bridgwater, A. V. (2003). Renewable fuels and chemicals by thermal processing of biomass. *Chemical Engineering Journal, 91*(2–3), 87–102. doi:10.1016/S1385-8947(02)00142-0.

Bridgwater, A. V. (2012). Review of fast pyrolysis of biomass and product upgrading. *Biomass and Bioenergy, 38*, 68–94. doi:10.1016/j.biombioe.2011.01.048.

Cascon, H., & Choudhary, S. (2013). Separation performance and stability of PVDF-co-HFP/alkylphosphonium dicyanamide ionic liquid gel-based membrane in pervaporative separation of 1-butanol. *Separation Science and Technology, 48*(11), 1616–1626. doi:10.1080/01496395.2012.762025.

Chapeaux, A., Simoni, L. D., Ronan, T. S., Stadtherr, M. A., & Brennecke, J. F. (2008). Extraction of alcohols from water with 1-hexyl-3-methylimidazolium bis(trifluoromethylsulfonyl)imide. *Green Chemistry, 10*(12), 1301–1306. doi:10.1039/b807675h.

Chapman, W. G., Gubbins, K. E., Jackson, G., & Radosz, M. (1989). SAFT: Equation-of-state solution model for associating fluids. *Fluid Phase Equilibria, 52*, 31–38. doi:10.1016/0378-3812(89)80308-5.

Chapman, W. G., Gubbins, K. E., Jackson, G., & Radosz, M. (1990). New reference equation of state for associating liquids. *Industrial & Engineering Chemistry Research, 29*(8), 1709–1721. doi:10.1021/ie00104a021.

Deb, K. (2001). *Multi-objective optimization using evolutionary algorithms*. Chichester, UK: John Wiley & Sons.

Diebold, J. P. (2000). A review of the chemical and physical mechanisms of the storage stability of fast pyrolysis bio-oils, NREL/SR-570-27613. http://www.nrel.gov/docs/fy00osti/27613.pdf.

Fischer, C. R., Klein-Marcuschamer, D., & Stephanopoulos, G. (2008). Selection and optimization of microbial hosts for biofuels production. *Metabolic Engineering, 10*(6), 295–304. doi:10.1016/j.ymben.2008.06.009.

Francisco, M., Arce, A., & Soto, A. (2010). Ionic liquids on desulfurization of fuel oils. *Fluid Phase Equilibria, 294*(12), 39–48. doi:10.1016/j.fluid.2009.12.020.

Freire, M. G., Carvalho, P. J., Gardas, R. L., Santos, L. M. N. B. F., Marrucho, I. M., & Coutinho, J. A. P. (2008). Solubility of water in tetradecyltrihexylphosphonium-based ionic liquids. *Journal of Chemical & Engineering Data, 53*(10), 2378–2382. doi:10.1021/je8002805.

Gao, H., Guo, C., Xing, J., Zhao, J., & Liu, H. (2010). Extraction and oxidative desulfurization of diesel fuel catalyzed by a Bronsted acidic ionic liquid at room temperature. *Green Chemistry, 12*(7), 1220–1224. doi:10.1039/C002108C.

Garbuz, S. V., Skopenko, V. V., Khavryuchenko, V. D., & Gerasimchuk, N. N. (1989). Quantum-chemical simulation of dicyanamide and tricyanomethanide ion solvation. *Theoretical and Experimental Chemistry, 25*(1), 83–86. doi:10.1007/bf00580304.

Garcia-Chavez, L. Y., Garsia, C. M., Schuur, B., & de Haan, A. B. (2012). Biobutanol recovery using nonfluorinated task-specific ionic liquids. *Industrial & Engineering Chemistry Research, 51*(24), 8293–8301. doi:10.1021/ie201855h.

Goldberg, D. E. (1989). *Genetic algorithms in search, optimization and machine learning.* Boston, MA: Addison-Wesley Publishing Company.

Gross, J., & Sadowski, G. (2001). Perturbed-chain SAFT: An equation of state based on a perturbation theory for chain molecules. *Industrial & Engineering Chemistry Research, 40*(4), 1244–1260. doi:10.1021/ie0003887.

Gross, J., & Sadowski, G. (2002). Application of the perturbed-chain SAFT equation of state to associating systems. *Industrial & Engineering Chemistry Research, 41*(22), 5510–5515. doi:10.1021/ie010954d.

Ha, S. H., Mai, N. L., & Koo, Y.-M. (2010). Butanol recovery from aqueous solution into ionic liquids by liquid–liquid extraction. *Process Biochemistry, 45*(12), 1899–1903. doi:10.1016/j.procbio.2010.03.030.

Hu, X., Mourant, D., Gunawan, R., Wu, L., Wang, Y., Lievens, C., Li, C.-Z. (2012). Production of value-added chemicals from bio-oil via acid catalysis coupled with liquid-liquid extraction. *RSC Advances, 2*(25), 9366–9370. doi:10.1039/C2RA21597G.

Huang, S. H., & Radosz, M. (1990). Equation of state for small, large, polydisperse, and associating molecules. *Industrial & Engineering Chemistry Research, 29*(11), 2284–2294. doi:10.1021/ie00107a014.

Huang, S. H., & Radosz, M. (1991). Equation of state for small, large, polydisperse, and associating molecules: extension to fluid mixtures. *Industrial & Engineering Chemistry Research, 30*(8), 1994–2005. doi:10.1021/ie00056a050.

Kubiczek, A., & Kamiński, W. (2013). Ionic liquids for the extraction of n-butanol from aqueous solutions. *Proceedings of ECOpole, 7*(1), 125–131.

Kuhlmann, E., Haumann, M., Jess, A., Seeberger, A., & Wasserscheid, P. (2009). Ionic liquids in refinery desulfurization: Comparison between biphasic and supported ionic liquid phase suspension processes. *Chemsuschem, 2*(10), 969–977. doi: 10.1002/cssc.200900142.

Lei, Z., Li, C., & Chen, B. (2003). Extractive distillation: A review. *Separation & Purification Reviews, 32*(2), 121–213. doi:10.1081/SPM-120026627.

Li, H., Jiang, X., Zhu, W., Lu, J., Shu, H., & Yan, Y. (2009). Deep oxidative desulfurization of fuel oils catalyzed by decatungstates in the ionic liquid of [Bmim]PF6. *Industrial & Engineering Chemistry Research, 48*(19), 9034–9039. doi:10.1021/ie900754f.

Mahfud, F. H., van Geel, F. P., Venderbosch, R. H., & Heeres, H. J. (2008). Acetic acid recovery from fast pyrolysis oil. An exploratory study on liquid-liquid reactive extraction using aliphatic tertiary amines. *Separation Science and Technology, 43*(11–12), 3056–3074. doi:10.1080/01496390802222509.

Matsumoto, M., Mochiduki, K., Fukunishi, K., & Kondo, K. (2004). Extraction of organic acids using imidazolium-based ionic liquids and their toxicity to *Lactobacillus rhamnosus*. *Separation and Purification Technology, 40*(1), 97–101. doi:10.1016/j.seppur.2004.01.009.

McCabe, W. L., Smith, J. C., & Harriott, P. (1993). *Unit Operations of Chemical Engineering*, 5th ed., McGraw Hill, Singapore.

Mohan, D., Pittman, C. U., & Steele, P. H. (2006). Pyrolysis of wood/biomass for bio-oil: A critical review. *Energy & Fuels, 20*(3), 848–889. doi:10.1021/ef0502397.

Nann, A., Held, C., & Sadowski, G. (2013). Liquid–liquid equilibria of 1-butanol/water/IL systems. *Industrial & Engineering Chemistry Research, 52*(51), 18472–18481. doi:10.1021/ie403246e.

Pei, Y., Wu, K., Wang, J., & Fan, J. (2008). Recovery of furfural from aqueous solution by ionic liquid based liquid–liquid extraction. *Separation Science and Technology, 43*(8), 2090–2102. doi:10.1080/01496390802064018.

Pereiro, A. B., & Rodriguez, A. (2009). Application of the ionic liquid Ammoeng 102 for aromatic/aliphatic hydrocarbon separation. *The Journal of Chemical Thermodynamics, 41*(8), 951–956. doi:10.1016/j.jct.2009.03.011.

Pilli, S. R., Banerjee, T., & Mohanty, K. (2014). Liquid–liquid equilibrium (LLE) data for ternary mixtures of [C4DMIM]—[PF6]+[PCP]+[water] and [C4DMIM][PF6]+[PA]+[water] at T = 298.15 K and p = 1 atm. *Fluid Phase Equilibria, 381*, 12–19. doi:10.1016/j.fluid.2014.08.004.

Plechkova, N. V., & Seddon, K. R. (2008). Applications of ionic liquids in the chemical industry. *Chemical Society Reviews, 37*(1), 123–150. doi:10.1039/B006677J.

Rasendra, C. B., Girisuta, B., Van deBovenkamp, H. H., Winkleman, J. G. M., Leijenhorst, E. J., Venderbosch, R. H., Windt, M., Meier, D., Heeres, H. J. (2011). Recovery of acetic acid from an aqueous pyrolysis oil phase by reactive extraction using tri-n-octylamine. *Chemical Engineering Journal*, 176–177, 244–252. doi:10.1016/j.cej.2011.08.082.

Renon, H., & Prausnitz, J. M. (1968). Local compositions in thermodynamic excess functions for liquid mixtures. *AIChE Journal, 14*(1), 135–144. doi:10.1002/aic.690140124.

Santiago, R. S., Santos, G. R., & Aznar, M. (2009). UNIQUAC correlation of liquid–liquid equilibrium in systems involving ionic liquids: The DFT–PCM approach. *Fluid Phase Equilibria, 278*(1–2), 54–61. doi:10.1016/j.fluid.2009.01.002.

Seader, J. D., & Henley, E. J. (2006). *Separation process principles*. New York: John Wiley & Sons.

Simoni, L. D., Chapeaux, A., Brennecke, J. F., & Stadtherr, M. A. (2010). Extraction of biofuels and biofeedstocks from aqueous solutions using ionic liquids. *Computers & Chemical Engineering, 34*(9), 1406–1412. doi:10.1016/j.compchemeng.2010.02.020.

Singh, M. K., Banerjee, T., & Khanna, A. (2005). Genetic algorithm to estimate interaction parameters of multicomponent systems for liquid–liquid equilibria. *Computers & Chemical Engineering, 29*(8), 1712–1719. doi:10.1016/j.compchemeng.2005.02.020.

Tariq, M., Serro, A. P., Mata, J. L., Saramago, B., Esperança, J. M. S. S., Canongia Lopes, J. N., & Rebelo, L. P. N. (2010). High-temperature surface tension and density measurements of 1-alkyl-3-methylimidazolium bistriflamide ionic liquids. *Fluid Phase Equilibria, 294*(1–2), 131–138. doi:10.1016/j.fluid.2010.02.020.

Vitasari, C. R., Meindersma, G. W., & de Haan, A. B. (2011). Water extraction of pyrolysis oil: The first step for the recovery of renewable chemicals. *Bioresource Technology, 102*(14), 7204–7210. doi:10.1016/j.biortech.2011.04.079.

Wauquier, J. P. (2000). *Petroleum Refining, Separation Processes*. Paris: Editions Technip.
Wei, Y. S., & Sadus, R. J. (2000). Equations of state for the calculation of fluid-phase equilibria. *AIChE Journal, 46*(1), 169–196. doi:10.1002/aic.690460119.
Wertheim, M. S. (1984a). Fluids with highly directional attractive forces. I. Statistical thermodynamics. *Journal of Statistical Physics, 35*(1), 19–34. doi:10.1007/bf01017362.
Wertheim, M. S. (1984b). Fluids with highly directional attractive forces. II. Thermodynamic perturbation theory and integral equations. *Journal of Statistical Physics, 35*(1), 35–47. doi:10.1007/bf01017363.
Wertheim, M. S. (1986a). Fluids with highly directional attractive forces. III. Multiple attraction sites. *Journal of Statistical Physics, 42*(3), 459–476. doi:10.1007/bf01127721.
Wertheim, M. S. (1986b). Fluids with highly directional attractive forces. IV. Equilibrium polymerization. *Journal of Statistical Physics, 42*(3), 477–492. doi:10.1007/bf01127722.
Wertheim, M. S. (1986c). Fluids of dimerizing hard spheres, and fluid mixtures of hard spheres and dispheres. *The Journal of Chemical Physics, 85*(5), 2929–2936. doi:10.1063/1.451002.
Wertheim, M. S. (1987). Thermodynamic perturbation theory of polymerization. *The Journal of Chemical Physics, 87*(12), 7323–7331. doi:10.1063/1.453326.
Wolbach, J. P., & Sandler, S. I. (1998). Using molecular orbital calculations to describe the phase behavior of cross-associating mixtures. *Industrial & Engineering Chemistry Research, 37*(8), 2917–2928. doi:10.1021/ie970781l.
Wu, Z. L., & Ondruschka, B. (2010). Ultrasound-assisted oxidative de-sulfurization of liquid fuels and its industrial application. *Ultrasonics Sonochemistry. 17*(6), 1027–1032. doi:10.1016/j.ultsonch.2009.11.005.
Yang, X. S. (2010). *Engineering optimization: An introduction with metaheuristic applications*. Hoboken, NJ: John Wiley & Sons.
Yang, X. S. (2014). *Nature-inspired optimization algorithms*. London: Elsevier.
Yu, J., Li, H., & Liu, H. (2006). Recovery of acetic acid over water by pervaporation with a combination of hydrophobic ionic liquids. *Chemical Engineering Communications, 193*(11), 1422–1430. doi: 10.1080/00986440500511478.

3

COSMO-SAC: A Predictive Model for Calculating Thermodynamic Properties on a-priori Basis

3.1 Liquid-Phase Thermodynamics

The prediction of properties of liquid-phase thermodynamics is conventionally performed using the Gibbs energy of solvation (ΔG^{solv}), which is merely the energy required to transfer a solute molecule from vacuum to a solvent phase. Two of the important properties computed from this route are the activity coefficient and the vapor pressures.

$$\ln\left(x_{i/s}\gamma_{i/s}\right)=\frac{\Delta G_{i/S}^{solv}-\Delta G_{i/i}^{solv}}{kT}=\frac{\Delta G_{i/S}^{*solv}-\Delta G_{i/i}^{*solv}}{kT}+\ln\left(\frac{V_{i/i}}{V_{i/S}}\right) \tag{3.1}$$

In the equation, $\Delta G_{i/S}^{*solv}$ is the Gibb's free energy of solvation of molecule i in solvent j, while $\Delta G_{i/i}^{*solv}$ refers to the term for molecule i in itself. Here, the term $\Delta G_{i/j}^{*,solv}$ refers to the non-translational contribution of the Gibb's free energy of solvation. This is related by Equations 3.2 and 3.3, as given below:

$$\Delta G_{i/j}^{*,solv}=\Delta G_{i/j}^{solv}-kT\ln\left(\frac{N\Lambda_i^3}{V_{i/j}}\right) \tag{3.2}$$

$$\Lambda_i=\frac{h}{\sqrt{2\pi m_i kT}} \tag{3.3}$$

Here, Equation 3.3 refers to the *De Broglie's* wavelength originating from the dual particle theory of material, where m_i is the mass of the particle, h is the Planck's constant (J.s) and k is the Boltzmann constant (J/K). In a similar manner, the vapor pressures can be predicted by Equation 3.4, where an ideal phase has been assumed.

$$\ln\left(\frac{P_i^{\text{vap}}V_{i/i}}{kT}\right) = \frac{\Delta G_{i/i}^{\text{solv}}}{kT} \tag{3.4}$$

It should be noted that the thermodynamic properties, namely partition coefficient, heat of vaporization and phase equilibria, are computed from $\gamma_{i/s}$ and P_i^{vap}. Hence, the objective of COSMO-SAC (Lin and Sandler, 2002) is to determine these properties from quantum chemical and statistical mechanical framework. The model is based on the original pioneering work of Klamt (Klamt, 1995; Klamt, Jonas, Bürger, & Lohrenz, 1998; Klamt & Schuurmann, 1993). The primary contribution from COSMO-SAC is that $\Delta G_{i/j}^{\text{solv}}$. $\Delta G_{i/j}^{\text{solv}}$ is mainly bifurcated from two contributors, namely electrostatic (ES) and van der Waals (vdW) force (Lin, Chang, Wang, Goddard, & Sandler, 2004; Wang, Lin, Chang, Goddard, & Sandler, 2006; Wang, Sandler, & Chen, 2007; Wang, Stubbs, Siepmann, & Sandler, 2005). Contributions such as those from vibrational, rotational and dipole–dipole moment are ignored. Let us write out the contribution for each part in Equation 3.5.

$$\begin{aligned}\Delta G_{i/S}^{*,\text{solv}} &= \Delta G_{i/S}^{*\text{ES}} + \Delta G_{i/S}^{*vdW} = \Delta G_{i/S}^{*\text{ES}} + \left[\Delta A_{i/S}^{*,vdW} + P\left(V_{i/S}^{\text{liq}} - V_i^{\text{Id.gas}}\right)\right] \\ &= \Delta G_{i/S}^{*\text{ES}} + \Delta A_{i/S}^{*,\text{disp}} + \Delta A_{i/S}^{*,\text{cav}} - kT\end{aligned} \tag{3.5}$$

$\Delta A_{i/S}^{*,vdW}$ is the Helmholtz energy of molecule *i* in the solvent. Assuming the fact that $V_i^{\text{Id.gas}} > V_{i/S}^{\text{liq}}$, we further write out the Helmholtz energy as a sum of two parts, namely $\Delta A_{i/S}^{*,\text{disp}}$ and $\Delta A_{i/S}^{*,\text{cav}}$. The former refers to the London Dispersion term. This is usually referred to as temporary attractive force, which occurs when electrons in two nearby atoms occupy positions such that they form temporary dipoles. This force is also called 'induced dipole–dipole' interaction. $\Delta A_{i/S}^{*,\text{cav}}$ refers to the amount of energy required to construct a cavity of component *i* within a solvent *S*. This primarily depends on the shape and size of the molecule.

The predictions of $\Delta G_{i/j}^{*\text{solv}}$ take a different pathway each for activity coefficients and vapor pressure. The activity coefficients are usually predicted from the difference of their properties in solvent and individual phase, as given below (Burnett, 2012):

$$\begin{aligned}\ln\left(x_{i/s}\gamma_{i/s}\right) &= \left(\frac{\Delta G_{i/S}^{*\text{ES}} + \Delta A_{i/S}^{*vdW} + \Delta A_{i/S}^{*\text{cav}} - kT}{kT}\right) \\ &\quad - \left(\frac{\Delta G_{i/i}^{*\text{ES}} + \Delta A_{i/i}^{*vdW} + \Delta A_{i/i}^{*\text{cav}} - kT}{kT}\right) + \ln\left(\frac{V_{i/i}}{V_{i/S}}\right)\end{aligned} \tag{3.6}$$

By rearranging the terms, we obtain:

$$= \left(\frac{\Delta G_{i/S}^{*ES} - \Delta G_{i/i}^{*ES}}{kT} \right) + \left(\frac{\Delta A_{i/S}^{*vdW} - \Delta A_{i/i}^{*vdW}}{kT} \right) + \left(\frac{\Delta A_{i/S}^{*cav} - \Delta A_{i/i}^{*cav}}{kT} \right) + \ln \left(\frac{V_{i/i}}{V_{i/S}} \right) \quad (3.7)$$

Both the phases considered are liquid; hence, the dispersion energies are almost similar and equal. In other words, their difference is fractionally small as compared with other terms. So, we neglect this contribution, that is, the second term of Equation 3.7. On similar lines, the last two terms reflect the differences in cavity formation and density. These are used through a combinatorial contribution, which is similar to that used in nonrandom two-liquid (NRTL) and UNI QUAsiChemical (UNIQUAC) models. Hence, the final expression takes the form (Equation 3.8):

$$= \left(\frac{\Delta G_{i/S}^{*ES} - \Delta G_{i/i}^{*ES}}{kT} \right) + \ln \left(\gamma_{i/S}^{\text{comb}} \right) \quad (3.8)$$

For the estimation of vapor pressures, we require the Gibb's free energy of solvation for a single phase. Hence, the dispersion terms do not cancel out; this invariably implies that the cavity formation term in Equation 5.8 (second term) is not straightforward. The Equation 3.8 takes the form:

$$\ln \left(\frac{P_i^{\text{vap}} V_{i/i}}{kT} \right) = \left(\frac{\Delta G_{i/i}^{*ES} + \Delta A_{i/i}^{*vdW} + \Delta A_{i/i}^{*cav} - kT}{kT} \right) \quad (3.9)$$

3.2 Estimation of Electrostatic Contribution*

Within the COSMO-SAC model, the largest contribution, as evident from Equation 3.8, is from the electrostatic contribution, which is due the inherent charges between molecules. The description of ΔG^{*ES} is a tedious job and is usually computed with a continuum solvation model, as defined in quantum mechanics. Here, the solute molecule is first modeled to be surrounded by a van der Waal-like segmented surface. Thereafter, it is immersed in a conductor with infinite dielectric constant, such that it totally screens the molecule's inherent electronic density. Density Functional Theory (DFT) is then applied on it to solve for the Hamiltonian, so as to determine the properties of each segment, including the charge and area associated with each segment. Thereafter, an ensemble-based theory along with a statistical mechanical framework is used, which relates the segment interactions to useful thermodynamic

* Sections 3.2–3.4 reprinted (adapted) from T. Banerjee, K. K. Verma and A. Khanna, Liquid–liquid equilibria for ionic liquid-based systems using COSMO-RS: Effect of cation and anion combination. *AIChEJ* 54, 1874–1885, 2008. Copyright Wiley-VCH Verlag GmbH & Co. KGaA. Reproduced with permission.

properties such as chemical potential. In a nutshell, this incorporates steps, namely geometry optimization of the molecule, construction of the vdW segmented surface, immersion of the molecule with the surface into a conductor and, finally, the computation of the segment properties via statistical mechanical framework. We shall now discuss all the steps in detail.

3.2.1 Geometry Optimization of the Molecule

The geometry for any given molecule is obtained by minimizing the total energy with respect to the atomic coordinates. This is usually performed by solving the Schrödinger equation, so as to predict the energies. Initially, the structures are drawn in some suitable freeware, such as Avogrado, Molden or GaussView, to name a view. Thereafter, the optimization is done in vacuum by giving a suitable level of theory (HF/DFT/Semi-Empirical) and a basis set in commercial packages such as Gaussian and TURBOMOLE. This is done in vacuum by using where after the end of optimization, an absence of negative frequencies indicates a global minimum. This step also gives us the optimized geometry, its partial charges and the energies of the molecule in vacuum. This optimized geometry is then used for all subsequent calculations. In so doing, the effect of any changes in geometry between vacuum and solvent is assumed negligible.

An interesting question may be: Should the geometry optimization be carried out in a known solvent? However, the choice of solvent may be then biased either towards high dielectric constants such as water or towards low dielectric constants such as hexane. Since neither of them is applicable, universal phenomena such as vacuum are chosen. This also reduces the computational time, as optimization in vacuum is faster. Moreover, reported experimental data (Pye, Ziegler, van Lenthe, & Louwen, 2009) suggest that optimization in vacuum provides better reproducibility.

3.2.2 Segmented Surface

The segment surface is created by a vdW-like surface, which is determined by the union of spheres drawn around each of the atoms in the geometry-optimized molecule. Here, each element corresponding to the particular atom is given a specific radius, with which the spheres are created. These radii, which are determined empirically, are called the 'COSMO radii' and these are roughly 20% more than their vdW radii. The COSMO radii for 11 elements, namely H, C, N, O, F, Si, P, S, Cl and Br, have been optimized by Klamt et al. (1998). For unknown elements, Klamt and Schuurmann (1993) have recommended a value 20% larger than the vdW radius. The optimized radii of the elements were obtained from 15,000 DFT calculations (Klamt et al., 1998). Given this extensive optimization, the commercial packages directly use these data in their database (Banerjee, Singh, & Khanna, 2006; Grensemann & Gmehling, 2005).

Once the molecular surface or the molecule is defined, it is then divided into tiny polyhedra (segments), each with its own position and area. Figure 3.1

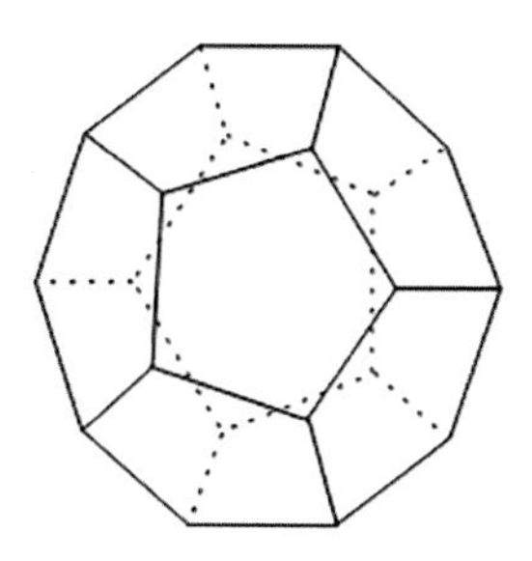
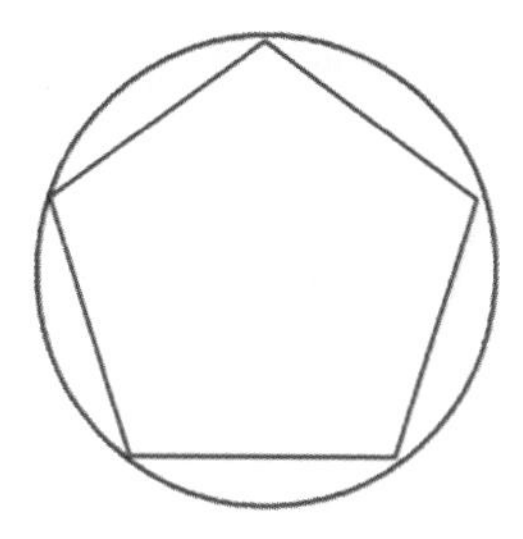

FIGURE 3.1
Sample surface segmentation into pentagons (left). The COSMO model approximates each of these by circles, with area equal to those of the original polygons (right).

shows a sample segmented sphere. Each segment can be either a pentagon or a hexagon, but for computational affordability, each is assumed circular, with a radius, $r_n = \sqrt{a_n/\pi}$.

In the initial DMOL3 implementation of COSMO, which was Klamt's original DFT software, the average segment size is 0.24 Å^2.

3.2.3 Determination of Segment Properties

After the segmented surface is obtained by the union of all the vdW spheres outlined around each atom, where each atom possesses a 'COSMO radii', the molecule is then immersed in a homogeneous conductor having infinite dielectric constant (Klamt, 2005). In such a scenario, the screening charges from the conductor migrate to the segmented surface to negate the underlying molecular charge distribution. This is physically modeled by placing a screening charge from the conductor on each segment. The total energy of solvation (E_{solv}) for the entire molecule is then determined from the interaction of the screening charges with the solute nuclei, the solute electronic density and within themselves. The segment properties are now calculated by defining the energy of solvation (E^{solv}), which solves the interaction of the segments with the solute nuclei (Z) (first term of Equation 3.1), solute electron density (second term of Equation 3.1) and interaction within the segments (third term of Equation 3.1), as given below. q_α represents the screening charge of αth segment.

$$4\pi\varepsilon_o E^{\text{solv}} = \sum_A \sum_\alpha Z_A q_\alpha B_{A\alpha} + \sum_\alpha q_\alpha C_\alpha + \frac{1}{2} \sum_\alpha \sum_\beta q_\alpha q_\beta D_{\alpha\beta} \tag{3.10}$$

where:

$$B_{A\alpha} = \frac{1}{|r_A - r_\alpha|}; \qquad C_\alpha = \int_V V_\alpha(r)\rho(r)dr$$

$$D_{A\alpha} = \frac{1}{|r_\alpha - r_\beta|}\ (\alpha \neq \beta); \quad D_{\alpha\alpha} = 1.07\sqrt{\frac{4\pi}{a_\alpha}}$$

Here B, C, D and r represent vector quantities and represent the contributions due to Coulomb's law. Z_α and r_α denote the molecular charge and position of nucleus A, respectively. V_α, possessing a unit of inverse of length, signifies the distance between the segment and the inherent charge density $\rho(r)$ of the molecule. $D_{\alpha\alpha}$ represents the energy required to provide a charge on each segment.

The expression for the self-energy of a segment was made by assuming the molecule as a sphere of radius $\boldsymbol{R}$. In such a situation, the total charge $\boldsymbol{Q}$ is distributed homogenously across its surface. Hence, the electrostatic energy of the sphere is given by

$$E = \frac{Q^2}{8\pi\varepsilon_0 R} \tag{3.11}$$

As per the COSMO protocol, the surface has to divide into n homogeneously similar segments. This implies that the energy then becomes the total of the sum of segment interaction energies and the segment self-energies; this is given by

$$\begin{aligned} E &= \sum_{\alpha=1}^{n}\sum_{\beta\geq\alpha}^{n} E_{\alpha\beta} + \sum_{\alpha=1}^{n} E_\alpha^{\text{self}} = \sum_{\alpha=1}^{n}\sum_{\beta\geq\alpha}^{n} \frac{q_\alpha q_\beta}{4\pi\varepsilon_0 r_{\alpha\beta}} + \sum_{\alpha=1}^{n} E_\alpha^{\text{self}} \\ &= \frac{n}{8\pi\varepsilon_0}\sum_{\beta=2}^{n}\frac{Q^2}{n^2 r_{1\beta}} + nE_\alpha^{\text{self}} \end{aligned} \tag{3.12}$$

Here, $E_{\alpha\beta}$ is the interaction energy between segments, while E_α^{self} is the self-energy of segment. The latter reflects the energy required to maintain a charge of a given density on a particular segment. q_α denotes the charge of segment α, while $r_{\alpha\beta}$ provides the distance between segments α and β. It is assumed that the area and charge of each segment are similar and the summations as above are exact. However, we still do not have an exact expression for E_α^{self}, since the energy of the sphere must be the same and will not depend on the distribution of charge on the sphere. An explicit way to determine the same is to an expression for can be obtained by equating Equations 3.11 and 3.12. Therefore, by rearranging the above equations, we obtain (Burnett, 2012):

$$\frac{Q^2}{8\pi\varepsilon_0 R} = \frac{n}{8\pi\varepsilon_0}\sum_{\beta=2}^{n}\frac{Q^2}{n^2 r_{1\beta}} + nE_\alpha^{\text{self}} \tag{3.13}$$

$$nE_\alpha^{\text{self}} = \frac{Q^2}{8\pi\varepsilon_0}\left[\frac{1}{R} - \frac{1}{n}\sum_{\beta=2}^{n}\frac{1}{r_{1\beta}}\right]$$

$$\Rightarrow \frac{Q^2}{8\pi\varepsilon_0 nR}\left[n - \sum_{\beta=2}^{n}\frac{1}{r_{1\beta}}\right]$$

By reorganizing the terms, we obtain:

$$\begin{aligned} E_\alpha^{\text{self}} &= \frac{Q^2}{8\pi\varepsilon_0 n^2 R}\sqrt{\frac{4\pi}{a_\alpha n}}\left[n - \sum_{\beta=2}^{n}\frac{1}{r_{1\beta}}\right] \\ &= \frac{q_\alpha^2}{8\pi\varepsilon_0}\sqrt{\frac{4\pi}{a_\alpha}}\sqrt{\frac{1}{n}}\left[n - \sum_{\beta=2}^{n}\frac{R}{r_{1\beta}}\right] \end{aligned} \tag{3.14}$$

Here, $R = \sqrt{A/4\pi}$ and is equal to $\sqrt{a_\alpha n/4\pi}$, A is the surface area of the sphere and $a_\alpha = \sqrt{A/n}$ is the area of each segment. Thus, in a nutshell, Equation 3.14 is exact only when the charges are homogenously placed on top of the sphere. However, as the charge and area are known from a COSMO calculation, this dependence is limited to only the last part of the equation, that is, $\sqrt{1/n}[n - \Sigma_{\beta=2}^{n} R/r_{1\beta}]$. The expression for this term is the highest (i.e. 1.250) at $n = 4$, while it steadily drops down to 1.078 at $n = 20$ and finally to 1.069 at $n = 60$. Hence, it has been recommended to keep this constant at $n = 20$ (i.e. 1.070) such that the final expression takes the form:

$$E_\alpha^{\text{self}} = \frac{1.07}{8\pi\varepsilon_0} q_\alpha^2 \sqrt{\frac{4\pi}{a_\alpha}} = \frac{0.15}{\varepsilon_0}\left(\frac{q_\alpha^2}{a_\alpha}\right) \tag{3.15}$$

The energy of solvation (E^{solv}) is then minimized with respect to the segments' charges (q_α), as given below:

$$\left(\frac{dE^{\text{solv}}}{dq_\alpha}\right) = \sum_A Z_A B_{A\alpha} + C_\alpha + \sum_\beta q_\beta D_{\alpha\beta} = 0 \tag{3.16}$$

As the molecule is placed on the continuum with infinite dielectric constant, the charges in the nearby region tend to cancel the electron density $\rho(\boldsymbol{r})$ of the solute molecule. This causes the molecule to polarize. Hence, this calculation needs to be performed in an iterative manner, so as to achieve consistency. At each step, the screening charge $\boldsymbol{q}$ is found for the respective $\rho(\boldsymbol{r})$, after which $\rho(\boldsymbol{r})$ is updated with new values of $\boldsymbol{q}$. This procedure, which utilizes

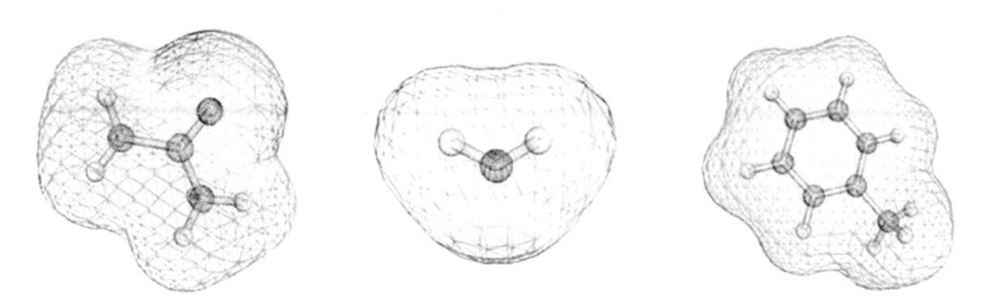

FIGURE 3.2
Gaussian03-generated COSMO surfaces for three molecules: Acetone (739 segments), water (292 segments) and toluene (1053 segments). Surfaces colored by screening charge density, where blue regions have negative screening charge values, green regions are neutral and red regions have positive values.

an iterative method, is called 'self-consistent field' approach. This process of generating and identifying the set of $\rho(\boldsymbol{r})$ and $\boldsymbol{q}$ that are self-consistent is called a 'conductor-like screening model' (COSMO).

In addition to $\boldsymbol{q}$ and area of the segments $\boldsymbol{a}$ and $\boldsymbol{r}$, it also provides the potential for each segment φ. The calculation also provides us with the total energy of the molecule in a conductor, which we will denote as ($\boldsymbol{E}_{\text{COSMO}}$). Sample screening charges for different molecules are depicted in Figure 3.2. It should be noted that the positive screening charges depict the negative regions of the molecule. Likewise, the negative screening charges represent the positive regions of the molecule. The screening charge distribution of the molecules is drawn by using the COSMO builder within TURBOMOLE. The package can also evaluate the segment potential for the segments, which we shall discuss in the ensuing section.

For the acetone COSMO file, *'nps'* represents the number of segments, which is 739. The values of *area* (Å^2) = 371.79 and *volume* (Å^3) = 561.92 represent the total cavity area and volume, respectively, within the conductor, where acetone molecule is placed. The next section denotes the position of all the 10 atoms, along with the total screening charge on each atom 'COSMOCharge' and the last column represents the division of this 'COSMOCharge' with the area corresponding to each atom, called 'area'. The last column is the most desirable quantity, as it reflects the screening charge density. It is interesting to note that except for oxygen atom 'O3' (COSMOCharge = 0.24046), all other atoms posses a negative charge. This follows the obvious trend, as negatively charged atoms possess positive screening charges, and vice versa. Further, the positive charge on the oxygen atom is then divided into a number of segments for which information is available in the concluding part of the COSMO file. For example, information for the segments of 'C1' and 'H9' atoms of acetone is depicted in Figure 3.3. The screening charges q of the solvent medium are usually calculated by a scaling q^*, so that $q = f(\varepsilon)\, q^*$, where $f(\varepsilon)$ is the scaling factor given by

$$f(\varepsilon) = \frac{\varepsilon - 1}{\varepsilon + 0.5} \tag{3.17}$$

```
Gaussian COSMO output
$cosmo_data
  fepsi =    1.00
  nps   =  739
  area  =  371.79
  volume=  561.92
$coord_rad
#atom   x   y   z   element  radius [A] area COSMOcharge sigma
 1  2.42938323473619 -1.15096746670694  0.00000200604284 6 2.00000
    18.67965   -0.01508   -0.00081
 2  0.00000790190236  0.32678868411764 -0.00001189287756 6 2.00000
    8.51712    -0.01501   -0.00176
 3  0.00011174943123  2.62986595127750  0.00000320803667 8 1.72000
    20.25649   0.24046    0.01187
 4 -2.42950234678886 -1.15080747183769 -0.00000101959499 6 2.00000
    18.67981   -0.01519   -0.00081
 5 -2.53466744490604 -2.36150522845060 -1.65106704759713 1 1.30000
    6.35161    -0.03540   -0.00557
 6 -4.01919024302989  0.12745236116114 -0.00078429598930 1 1.30000
    6.28553   -0.02677    -0.00426
 7 -2.53529356852293 -2.36025933097624  1.65192971664273 1 1.30000
    6.35202   -0.03540    -0.00557
 8  2.53468635185488 -2.36133231522858  1.65128747790871 1 1.30000
    6.35152    -0.03540   -0.00557
 9  4.01918033927670  0.12718985822214  0.00049579434371 1 1.30000
    6.28566    -0.02682   -0.00427
10  2.53505783077930 -2.36055542838593 -1.65182187102383 1 1.30000
    6.35194    -0.03540   -0.00557
$coord_car
!BIOSYM archive 3
PBC=OFF
coordinates from GAUSSIAN/COSMO calculation
!DATE
C1    1.285574238 -0.609065751  0.000001062 GAUS 1   C   C    0.0000
C2    0.000004182  0.172929124 -0.000006293 GAUS 1   C   C    0.0000
O3    0.000059135  1.391665122  0.000001698 GAUS 1   O   O    0.0000
C4   -1.285637269 -0.608981085 -0.000000540 GAUS 1   C   C    0.0000
H5   -1.341288242 -1.249654744 -0.873707051 GAUS 1   H   H    0.0000
H6   -2.126863872  0.067444885 -0.000415032 GAUS 1   H   H    0.0000
H7   -1.341619573 -1.248995444  0.874163556 GAUS 1   H   H    0.0000
H8    1.341298248 -1.249563242  0.873823698 GAUS 1   H   H    0.0000
H9    2.126858632  0.067305974  0.000262363 GAUS 1   H   H    0.0000
H10   1.341494826 -1.249152132 -0.874106486 GAUS 1   H   H    0.0000
end
end
$screening_charge
  cosmo      =  -0.017099
  correction =   0.017099
  total      =   0.000000
$cosmo_energy
  Total energy corrected [a.u.]  =     -193.0450861463
  Dielectric energy corr. [a.u.] =     -0.0105147256
```

FIGURE 3.3
An excerpt of a COSMO file, as generated from Gaussian03 version C.02 for acetone molecule.

(Continued)

```
$segment_information
# n              - segment number
# atom           - atom associated with segment n
# position       - segment coordinates [a.u.]
# charge         - segment charge (corrected)
# area           - segment area [A**2]
# potential      - solute potential on segment (A length scale)
#
#  n atom  position (X, Y, Z)    charge   area   charge/area  potential
#
#
1 9 3.724444694 2.287044652 1.133305480 -0.000211555 0.085032594
-0.002487925   -0.037535002
2 9 4.806433344 0.909518246 2.192137785 -0.000120711 0.011682637
-0.010332516
---------------------------------------------------------------------------------------
---------------------------------------------------------------------------------------
  739    1    4.828863754    0.396052515    -2.476584729
-0.000064470    0.128565395    -0.000501458
-0.008333038
```

FIGURE 3.3 (Continued)
An excerpt of a COSMO file, as generated from Gaussian03 version C.02 for acetone molecule.

Thus, a simpler boundary condition of a conductor appears in the above equation and also in Figure 3.3 (*fepsi* = 1.00). It should be noted that a conductor has an infinite supply of charge, so as to screen the entire solute molecule. Hence, there is a point of thought that the selection of solvent as a conductor can alter the charges, as evident in Figure 3.2. Therefore, a decision has to be arrived for the selection of a solvent in such a manner so that it is able to screen the entire solute component. This brings out to the point that it should possess sufficient charge, and thus, for obvious reason, it has to be a perfect conductor. Here, the COSMO approximation is exact in the limit of $\varepsilon = \infty$ and is within 0.5% accuracy for strong dielectrics such as water with a permeability of ($\varepsilon = 80$). Even for a lower dielectric limit of solvents such as $\varepsilon = 2$, COSMO coincides with the exact dielectric model within 10%. Hence, the COSMO approach is becoming a standard Continuum Solvation Model (CSM) in quantum chemical codes.

It should be noted that the self-consistent field (SCF) calculation or the COSMO scheme has to be done once for each component. Once these charges are obtained, then a framework has to be created so as to restore these charges when placed in an actual solvent of a known dielectric constant. This is where the derivation of chemical potentials and activity coefficients is performed. Our aim will then be to create a repository of COSMO files, which will include ions, solutes, solvents and inorganic compounds, subject to their dissociation constants. A sample description of an input for generating the COSMO file is given in Figure 3.4.

The different terms and their meaning are as follows:

'% mem = 540 MW' refers to the total amount of internal memory required to perform this calculation. The route section with '# P BVP86/SVP/DGA1'

```
%chk=acetone
#P BVP86/svp/DGA1 scf=tight

User defined Comment Line

0  1          (Charge and Multiplicity)

Z-Matrix

--link1--
%chk=acetone
#P BVP86/svp/DGA1 scf=(tight,novaracc) SCRF=COSMORS guess=read geom=checkpoint

User defined Comment line

0  1                    (Charge and Multiplicity)
acetone.cosmo           (User defined COSMO file)
```

FIGURE 3.4
Input file for COSMO file generation in Gaussian03.

indicates the following form 'DFT Theory/Basis Set/optional'. So, the form of functional used for DFT is the P BVP86, while the basis set used is SVP or Split Valence Polarized. The orbital coefficients for SVP can be obtained by the *GFprint* and *GFInput* commands in Gaussian03. The density gradient approximation or 'optional' used is DGA1. It expands the electron density of the atoms in the form of atom-centered functions for saving computational time. It only integrates the coulomb interaction in place of all the electron integrals. The 'SCF = tight' indicates a full convergence of energy.

3.2.4 Computation of the Restoring Free Energy $\left(\Delta G_{i/s}^{*es}\right)$

Here, the contribution to the restoring free energy can be divided into two parts, namely ($\Delta G_{i/s}^{*es}$) and (ΔG_i^{*IS}), where the former implies the electrostatic energy due to the component in the solution and the latter explains the change in energy when the molecule is taken from a conductor to a solvent. These are mathematically defined as below:

$$\begin{aligned}\Delta G_{i/s}^{*es} &= \Delta G_{i/s}^{*\text{res}} + \Delta G_i^{*IS} \\ \Delta G_i^{*IS} &= E_{\text{COSMO}} - E_{\text{vacuo}}\end{aligned} \tag{3.18}$$

Both the energies in COSMO conductor and vacuum optimization can be obtained as per the procedures mentioned in Sections 3.2.2 and 3.2.1, respectively. For example, the total energy in the conductor, as per Figure 3.3, is

'Total energy corrected [a.u.] = –193.0450861463'. $\Delta G_{i/s}^{*\text{res}}$ represents the restoring free energy for the movement of the solute from the conductor to real solvent. For a real conductor, $\Delta G_{i/s}^{*\text{res}} = 0$, which implies that $\Delta G_{i/s}^{*es} = \Delta G_i^{*IS}$. Hence, the major portion from now on will be the Segment Activity Coefficient (SAC) approach, which will involve the derivation of segment potential, as discussed in the next section.

3.2.5 Segment Chemical Potential

Based on statistical thermodynamics, COSMO-SAC assumes the ensemble of molecules by an ensemble of independently interacting segments (Lin and Sandler, 2002; Burnett, 2012). Hence, for any mixture, segments from each component are mixed in proportion with the desired component mole fractions. The resulting interaction is then captured by the energy required to bring them from an infinite point in space. Let us consider an ensemble of k different types of molecules, each having n_i segments. Hence, the total number of segments is

$$n_c = \sum_{i=1}^{k} n_i$$
$$n^{\text{tot}} = \sum_{i=1}^{k} b_i n_i = \sum_{\alpha=1}^{n_c} \eta_\alpha \tag{3.19}$$

Here, b represents the total number of molecules of type i and η_α represents the total number of segments of type α. For a single component, this reduces to $\eta_\alpha = \eta_\beta = b_i$. Assuming a constant geometry, where all the segments are allowed to move freely, implies that the movements of the segments are not dependent on neighboring segments. This ensemble then takes a form of mixture of independently interacting segments. We can thus propose any such formation as below (Burnett, 2012):

$$\begin{bmatrix} 2\eta_{1,1} + \eta_{1,2} + \eta_{1,3} + \ldots\ldots.. \quad + \eta_{1,n_c} \\ \eta_{2,1} + 2\eta_{2,2} + \eta_{1,3} + \ldots\ldots.. \quad + \eta_{2,n_c} \\ \ldots\ldots \\ \ldots\ldots. \\ \eta_{n_c,1} + \eta_{n_c,2} + \eta_{n_c,2} + \ldots\ldots.. + 2\eta_{n_c,n_c} \end{bmatrix} = \begin{bmatrix} \eta_1 \\ \eta_2 \\ .. \\ .. \\ \eta_{n_c} \end{bmatrix} \tag{3.20}$$

Each row in the above matrix notation denotes the interaction of the first with the remaining segments (same component or of other component). Hence, the total neighbors will be of type '1' and will be η_1 such that $\eta_{\text{total}} = \sum_{i=1}^{n_c} \eta_i$. So, from the same logic, the number of pairs in the mixture will be

$$n_{\text{pairs}} = \sum_{\alpha=1}^{n_c} \eta_{\alpha\alpha} + \frac{1}{2}\sum_{\alpha=1}^{n_c}\sum_{\beta=1}^{n_c} \eta_{\alpha\beta} = \frac{1}{2}\sum_{\alpha=1}^{n_c} \eta_{\alpha} = \frac{n_{\text{total}}}{2} \tag{3.21}$$

Thus, the probability of obtaining a segment pair $\eta_{\alpha\beta}$ will be

$$p_{\alpha\beta} = \frac{\langle \eta_{\alpha\beta} \rangle}{n_{\text{pairs}}} \tag{3.22}$$

Here, $\langle \eta_{\alpha\beta} \rangle$ is an ensemble averaged value approximated from a canonical ensemble. Hence, the whole mixture can be written in terms of a partition function $Q(N, V, T)$ with a Boltzmann exponential term (see Appendix I for canonical partition function). This is given as below

$$p_{\alpha\beta} = \omega_{\alpha\beta} \exp\left(\frac{-E_{\alpha\beta}}{kT}\right) \frac{Q(n_{\text{pairs}} - \eta_{\alpha\beta}, V, T)}{Q(n_{\text{pairs}}, V, T)} \tag{3.23}$$

Here, Q and Q' represent the partition functions of the mixture with and without including the segment pair, respectively. Here, $E_{\alpha\beta}$ represents the interaction between the segment pair α and β. Thus, when $\alpha = \beta$, then $\omega_{\alpha\beta} = 1$, since both the segments become indistinguishable. The value of the degeneracy or $\omega_{\alpha\beta}$ becomes 2 when $\alpha \# \beta$. Equation 3.23 takes the following form when average probability is defined:

$$p_{\alpha\beta} = \omega_{\alpha\beta} \exp\left(\frac{-E_{\alpha\beta}}{kT}\right) \frac{Q(n_{\text{total}} - \chi_{\alpha} - \chi_{\beta}, V, T)}{Q(n_{\text{total}}, V, T)} \tag{3.24}$$

The difference between Equations 3.23 and 3.24 is that they denote a segment pair and an individual segment, respectively. It is a known fact the Helmholtz energy for the ensemble having the two segments χ_{α}, χ_{β} takes the form:

$$A(n_{\text{total}} - \chi_{\alpha} - \chi_{\beta}, V, T) = -kT \ln Q(n_{\text{total}} - \chi_{\alpha} - \chi_{\beta}, V, T) \tag{3.25}$$

From statistical thermodynamics, we know:

$$\left(\frac{\partial G}{\partial N_i}\right)_{T,P,N_{j\neq i}} = \left(\frac{\partial A}{\partial N_i}\right)_{T,V,N_{j\neq i}} = -kT\left(\frac{\partial \ln Q}{\partial N_i}\right)_{T,V,N_{j\neq i}} = \mu_i(\text{chemical potential}) \tag{3.26}$$

Hence, Equation 3.25 takes the form:

$$A(n_{\text{total}} - \chi_{\alpha} - \chi_{\beta}, V, T) = A(n_{\text{total}}) - \mu_{\alpha} - \mu_{\beta} \tag{3.27}$$

Here, μ_α, μ_β are the chemical potentials for the two segments α and β, respectively. Hence, we have:

$$\frac{Q(n_{\text{total}}-\chi_\alpha-\chi_\beta,V,T)}{Q(n_{\text{total}},V,T)}=\frac{\exp\left(\dfrac{-A\left(n_{\text{total}}-\chi_\alpha-\chi_\beta,V,T\right)}{kT}\right)}{\exp\left(\dfrac{-A\left(n_{\text{total}},V,T\right)}{kT}\right)} \tag{3.28}$$

$$\Rightarrow \exp\left(\frac{-A\left(n_{\text{total}},V,T\right)-A\left(n_{\text{total}}-\chi_\alpha-\chi_\beta,V,T\right)}{kT}\right)$$

$$\Rightarrow \exp\left(\frac{\mu_\alpha+\mu_\beta}{kT}\right)$$

Equation 3.24 then takes the form:

$$p_{\alpha\beta}=\omega_{\alpha\beta}\exp\left(\frac{-E_{\alpha\beta}}{kT}\right)\exp\left(\frac{\mu_\alpha+\mu_\beta}{kT}\right) \tag{3.29}$$

The non-translational part of the chemical potential is given as

$$\mu^*=\mu-kT\ln\left(\rho\wedge^3\right) \tag{3.30}$$

This implies that the second part is the translational contribution for chemical potential. So, for ideal gases, the non-translational part is usually zero, or

$$\mu^*=\mu-kT\ln\left(\rho\wedge^3\right)\ \therefore\ \mu^*\Rightarrow\mu=kT\ln\left(\rho\wedge^3\right) \tag{3.31}$$

The final form of Equation 3.29 takes the form:

$$p_{\alpha\beta}=\omega_{\alpha\beta}\exp\left(\frac{-E_{\alpha\beta}}{kT}\right)\exp\left(\frac{\mu_\alpha^*+\mu_\beta^*}{kT}\right) \tag{3.32}$$

For obtaining the average number of ensemble pairs, we thus use the relation as below:

$$\chi_{\alpha\beta}=n_{\text{pairs}}p_{\alpha\beta}=n_{\text{pairs}}\omega_{\alpha\beta}\exp\left(\frac{-E_{\alpha\beta}}{kT}\right)\exp\left(\frac{\mu_\alpha^*+\mu_\beta^*}{kT}\right) \tag{3.33}$$

The total number of degrees of freedom for such a system can be shown to be equal to $n^c/2(n^c+3)$; hence, we require the same number of equations so as to solve them and get a unique solution. Rewriting Equation 3.22 by diving it with $\omega_{\alpha\beta}$, we obtain:

$$\sum_{\beta=1}^{n_c} \frac{p_{\alpha\beta}}{\omega_{\alpha\beta}} = \left[\frac{\eta_{\alpha 1}}{2} + \frac{\eta_{\alpha 2}}{1} ..... + \frac{\eta_{\alpha\varepsilon}}{1} + ... + \frac{\eta_{\alpha n_c}}{2} \right] \frac{1}{n_{\text{pairs}}} \tag{3.34}$$

Here, the total probability is divided with the degeneracy and finally with total number of segment pairs. In a way, the above expression represents all possible probabilities of segment α with remaining segments across all the components *c*.

$$= \left[\eta_{\alpha 1} + \eta_{\alpha 2} + ...2\eta_{\alpha\alpha} + ... + \eta_{\alpha n_c} \right] \frac{1}{n_{\text{total}}} \tag{3.35}$$

Equation 3.35 points out to the fact that the total interaction with the remaining segments merely points out to the total number of pairs with segment α. Hence, we obtain *n*:

$$\Rightarrow \frac{\eta_\alpha}{n_{\text{total}}} = x_\alpha \tag{3.36}$$

Further, Equation 3.36 points out to the fact that a probability of locating a particular segment pair in the ensemble is merely locating a single pair or the segment mole fraction x_α.When applied to Equation 3.29, it takes the form:

$$\sum_{\beta=1}^{n_c} \frac{p_{\alpha\beta}}{\omega_{\alpha\beta}} = \sum_{\beta=1}^{n_c} \frac{\omega_{\alpha\beta}}{\omega_{\alpha\beta}} \exp\left(\frac{-E_{\alpha\beta}}{kT} \right) \exp\left(\frac{\mu_\alpha^* + \mu_\beta^*}{kT} \right) \tag{3.37}$$

$$\Rightarrow \exp\left(\frac{-\mu_\alpha^*}{kT} \right) \sum_{\beta=1}^{n_c} \exp\left(\frac{-E_{\alpha\beta}}{kT} \right) \exp\left(\frac{\mu_\beta^*}{kT} \right) \tag{3.38}$$

By combining the above two expressions, we obtain:

$$x_\alpha \Rightarrow \exp\left(\frac{-\mu_\alpha^*}{kT} \right) \sum_{\beta=1}^{n_c} \exp\left(\frac{-E_{\alpha\beta}}{kT} \right) \exp\left(\frac{\mu_\beta^*}{kT} \right) \tag{3.39}$$

Rearranging the equation provides the following form:

$$\exp\left(\frac{-\mu_\alpha^*}{kT} \right) \Rightarrow x_\alpha \left(\sum_{\beta=1}^{n_c} \exp\left(\frac{-E_{\alpha\beta}}{kT} \right) \exp\left(\frac{\mu_\beta^*}{kT} \right) \right)^{-1} \tag{3.40}$$

This can be rewritten as

$$\Gamma_{\alpha}^{*} = x_{\alpha}\left(\sum_{\beta=1}^{n_c}\Gamma_{\beta}^{*}\,\tau_{\alpha\beta}\right)^{-1} \tag{3.41}$$

where:

$$\Gamma_{\beta}^{*} = \exp\left(\frac{-\mu_{\alpha}^{*}}{kT}\right) \quad \text{and} \quad \tau_{\alpha\beta} = \exp\left(\frac{-E_{\alpha\beta}}{kT}\right) \tag{3.42}$$

Thus, the equation is implicit in nature, as μ_{α}^{*} appears on both sides of the equation. Hence, this needs to be solved iteratively. The $E_{\alpha\beta}$ possesses the form as given below:

$$\begin{aligned} E_{\alpha\beta} &= f_{\text{pol}}\frac{0.15a_{\text{eff}}^{3/2}}{\varepsilon_0}\left(\sigma_{\alpha}+\sigma_{\beta}\right)^2 \\ &= c_{es}\left(\sigma_{\alpha}+\sigma_{\beta}\right)^2 \end{aligned} \tag{3.43}$$

Here, a_{eff} is the area of the effective segment and is an adjustable parameter. $\sigma_{\alpha} = q_{\alpha}/a_{\alpha}$ is charge density of segments. Here, c_{es} is a constant and is equal to $f_{\text{pol}}\,0.15a_{\text{eff}}^{3/2}/\varepsilon_0$. The equation above reflects the self-consistency approximation, as discussed earlier. Here, all the segments are assumed to be of the same size as per 'area of the effective segment'; hence, the segment pair itself can be assumed to be an individual identity, with a total charge as the sum of the two charge segments. As the two molecules interact with each other, their electronic densities and subsequently, the wave function get disturbed. To capture this effect, the factor f_{pol} is defined.

Now, we shall move to the most important aspect of thermodynamics, that is, the derivation of the chemical potential from the information available from the segment charge densities. In such a scenario, we shall assume the entire mixture of N molecules to be divided into number of segments, $n_1, n_2 \ldots n_n$. Assuming that we have large number of molecules implies that removal of a single molecule or segment will define another system. An analogy can be devised from an NPT ensemble, where a deduction of a particular molecule (mol_i) or a segment (seg_{α}) gives a pathway for calculating the chemical potential of the particular molecule or a segment, respectively.

$$G\left(N-\text{mol}_i,P,T\right) = G\left(N,P,T\right) - \mu_{\text{mol},i}$$

Here, $\mu_{\text{mol},i}$ refers to the chemical potential of the *ith* molecule from the mixture. Now, if we consider the same ensemble but in terms of segments (seg_{α}), we obtain (Burnett, 2012):

$$G(N-\text{seg}_\alpha,P,T)=G(N,P,T)-\mu_{\text{seg},\alpha} \tag{3.44}$$

So, by summing up the contributions from all the segments, we obtain:

$$\mu_i=\sum_{\alpha=1}^{N}\mu_\alpha \tag{3.45}$$

So, the total energy or the restoring free energy required will be equal to the sum of all segments in a mixture S. This is given as below:

$$\Delta G_{i/S}^{*\text{res}}=\sum_{\alpha=1}^{n_i}\mu_{\alpha/S}^{*} \tag{3.46}$$

However, Equation 3.44 shall also be valid when the solvent itself is a perfect conductor. This will be applicable for the following condition:

$$\Delta G_{i/S}^{*\text{res}}=\sum_{\alpha=1}^{n_i}n_i^{\text{eff}}x_{\alpha/i}\left(\mu_{\alpha/S}^{*}-kT\ln x_{\alpha/S}\right) \tag{3.47}$$

Here, n_i^{eff} denotes the effective number of segments of component I, while $n_i^{\text{eff}}x_{\alpha/i}$ gives us the effective number of segments of molecule I of type α. So, when it becomes a perfect conductor, then Equation 3.47 takes the form:

$$\begin{gathered}\Delta G_{i/S}^{*\text{res}}=0\\ \sum_{\alpha=1}^{n_i}n_i^{\text{eff}}x_{\alpha/i}\left(\mu_{\alpha/S}^{*}-kT\ln x_{\alpha/S}\right)=0\end{gathered} \tag{3.48}$$

Thus, we have to subtract the term $kT\ln x_{\alpha/S}$ from the chemical potential to obtain the non-translational contribution.

$$\begin{gathered}\mu_{\alpha/S}^{*}-kT\ln x_{\alpha/S}=0\\ \Rightarrow\mu_{\alpha/S}^{*}=kT\ln x_{\alpha/S}\end{gathered} \tag{3.49}$$

This merely reflects the fact that for a perfect conductor, the restoring free energy is zero and the chemical potential is due to its translational contribution. Equation 3.47 is based on the premises that all segment interactions are possible. In order to assume such information, we need to neglect the three-dimensional segment information. Further, no attempt has been made to consider the steric hindrance or the long-range liquid structure. This may be termed the most limiting assumption in the model. On the contrary, this assumption only makes the model simplistic in nature, as it eliminates the need to explicitly sample the full configuration space of the entire ensemble.

In the stated COSMO calculation, the surface of the molecule is dissected in small segments and screening charges are determined for each segment, such that the net potential everywhere at the surface is zero (perfect screening). These charges are averaged in order to make the predictions uniform and also to capture the different shapes and sizes of the segments uniformly. Thus, to obtain apparent screening charges to be used in COSMO-SAC model, the averaging algorithm is proposed as below:

$$\sigma_m = \frac{\sum_n \sigma_n^* \frac{r_n^2 r_{\text{eff}}^2}{r_n^2 + r_{\text{eff}}^2} \exp\left(-\frac{d_{mn}^2}{r_n^2 + r_{\text{eff}}^2}\right)}{\sum_n \frac{r_n^2 r_{\text{eff}}^2}{r_n^2 + r_{\text{eff}}^2} \exp\left(-\frac{d_{mn}^2}{r_n^2 + r_{\text{eff}}^2}\right)} \tag{3.50}$$

where:
$r_{\text{eff}} = \sqrt{a_{\text{eff}}/\pi}$ is the radius of the standard surface segment
$r_n = \sqrt{a_n/\pi}$ is the radius of the segment n
d_{mn} is the distance between segments m and n

The averaging of the screening charge densities, therefore, creates a new contribution to the overall Gibb's free energy. This is provided as below:

$$\Delta G_{i/s}^{*es} = \Delta G_{i/s}^{*\text{res}} + \Delta G_i^{*IS} + \Delta G_i^{*cc} \tag{3.51}$$

Here, the third term on the right-hand side is the Gibb's free energy change occurring due to the averaging process.

$$\Delta G_{cc(i)}^{*} = f_{\text{pol}}^{1/2}\left[E_i^{\text{diel}}(q) - E_i^{\text{diel}}(q^*)\right] \tag{3.52}$$

Thus, the first term, that is, the restoring free energy, is a function of the solvent, while the remaining terms are dependent on the molecule configuration. Here, E^{diel} is given by

$$E_i^{\text{diel}}(q) = \frac{1}{2}\sum_\mu \phi_\mu q_\mu \tag{3.53}$$

$E_{\text{diel}}(q^*)$ is directly obtained from COSMO output file and is the energy needed for dielectric polarization, that is, the screening charge distribution. The factor has a constant value of $f_{\text{pol}} = 0.6917$. These two free energies, namely the second and third terms of Equation 3.51, are pure species properties. So, they have the same values in ideal gas phase (i.e. in vacuum) and in the solvent. Mixture properties are thus obtained by subtracting free energy values of mixture and vacuum phase respectively. This finally leads to the cancellation for mixture property calculation. This is the reason that they are not included in the final expression in solvation free energy.

3.2.6 Sigma Profiles of Simple Compounds

The three-dimensional screening charge density distributions, which are obtained as per the procedure given in Figure 3.3, are first averaged using Equation 3.50. This gives us charges ranging from –0.03 e/Å^2 to +0.03 e/Å^2. These segment charge densities are grouped using a histogram, which is commonly known as σ-profile ($p(\sigma)$). This essentially is the probability of locating a surface segment with screening charge density σ by the expression:

$$p_i(\sigma) = \frac{A_i(\sigma)}{A} \tag{3.54}$$

where:

$A_i(\sigma)$ is the surface area with a charge density of value σ
a_i is the total surface area of species i

For a mixture, the σ-profile is determined from the area-weighted average of contributions from all the components present in the mixture. This is given as

$$p_s(\sigma) = \frac{\sum_i x_i A_i p_i(\sigma)}{\sum_i x_i A_i} \tag{3.55}$$

Since the current theme of this book is on ionic liquids (ILs), we shall introduce sigma profiles for simple components and then move towards ILs. It should be noted that computation of segment potential is quite a cumbersome process, concerning the number of segments a molecule possesses. A different and easier approach would be to use the histogram known as σ-profile or $p(\sigma)$. In such a scenario, Equation 3.50 takes the form:

$$\exp\left(\frac{-\mu_S^*(\sigma_\alpha)}{kT}\right) = p_s(\sigma_\alpha)\left\{\sum_{\beta=1}^{N_{\text{bins}}} \exp\left(\frac{-E_{\alpha\beta}}{kT}\right)\exp\left(\frac{\mu_S^*(\sigma_\alpha)}{kT}\right)\right\} \tag{3.56}$$

$$\Delta G_{i/s}^{*\text{res}} = n_i\left\{\sum_{\beta=1}^{N_{\text{bins}}} p_i(\sigma_\alpha)\left\{\mu_S^*(\sigma_\alpha) - kt\ln p_s(\sigma_\alpha)\right\}\right\} \tag{3.57}$$

Here, σ_α refers to the charge density within the histogram. Now, it is known that the histograms lie between –0.03 e/Å^2 and +0.03 e/Å^2. So, if we allocate each histogram with a value of 0.001 e/Å^2, we shall then reduce the whole sigma profile to 61 histograms. Hence, N_{bins} takes the number 61. In the above altered equations, $x_{\alpha/i}$ and $x_{\alpha/S}$ are invariably replaced by $p_i(\sigma_\alpha)$ and $p_S(\sigma_\alpha)$, respectively. The latter two determine the sigma profiles of the component in itself (Equation 3.54) and in the mixture (Equation 3.55), respectively.

However, x_α is segment mole fraction, while $p_S(\sigma_\alpha)$ is based on the segment area fraction. Notwithstanding as Equation 3.47 is not derived, the difference is usually very small when computing $\Delta G_{i/S}^{*res}$ from $\mu_{\alpha/S}^{*}$.

It is interesting to note that the exchange energy (E_{pair}) contains contributions from electrostatic interactions. One of them is referred as misfit energy E_{mf}, which is the energy computed due to the difference in shape and size between a pair of segments. Another contribution, the hydrogen bonding energy E_{hb}, occurs when an electropositive atom (hydrogen) forms additional bonds with electronegative atoms such as oxygen, nitrogen and fluorine. The expressions of the two terms are given by Equations 3.58 and 3.59, respectively.

$$E_{mf} = (\alpha'/2)(\sigma_\alpha + \sigma_\beta)^2 \tag{3.58}$$

$$E_{hb} = c_{hb}\min\left[0, \max(0, \sigma_{acc} - \sigma_{hb})\min(0, \sigma_{don} + \sigma_{hb})\right] \tag{3.59}$$

$$E_{pair}(\sigma_\alpha, \sigma_\beta) = E_{mf} + E_{hb} \tag{3.60}$$

Where, α' is a constant for the misfit energy and is calculated from the surface area of a standard segment.

$$\alpha' = \frac{0.64 \times 0.3 \times a_{eff}^{3/2}}{\varepsilon_0}$$

$$e_0 = 2.395 \times 10^{-4}\ (e^2\ mol)/(kcal\ Å)$$

Here, c_{hb} is a constant for the hydrogen bonding interactions, while σ_{hb} is cutoff value for hydrogen bonding interactions. When Equation 3.59 is used, all possible pairs are first grouped. Now, each pair is dealt with separately, having screening charge densities as σ_α and σ_β. Then, σ_{acc} and σ_{don} are allocated the larger and smaller values of σ_m and σ_n. Here max and min indicate the larger and smaller values of their arguments, respectively. Under this definition, the hydrogen bonding contribution is non-zero only if one segment has a negative charge density less than $|\sigma_{hb}|$ and the other has a positive charge density greater than σ_{hb}. In such a case, the contribution is limited to segment pairs of opposite charge possessing larger $|\sigma_{hb}|$ magnitudes. This contribution will also increase higher difference in charges densities between the pair of segments. The SAC then takes up the form:

$$\ln\Gamma_s(\sigma_\alpha) = -\ln\left\{\sum_{\sigma_n} p_s(\sigma_\alpha)\Gamma_s(\sigma_\alpha)\exp\left[-\frac{E^{pair}(\sigma_\alpha, \sigma_\beta)}{kT}\right]\right\} \tag{3.61}$$

This equation is then solved iteratively. In the next section, we will focus our attention on the sigma profile and potential of the commonly used cations and anions.

3.2.7 Sigma Profiles and Potentials of Ionic Liquids

The sigma profile and potential of cations and anions are obtained using Equations 3.55 and 3.56, respectively (Figures 3.5 and 3.6). For such an exercise, we have chosen a common cation, that is, 1-alkyl-3-methylimidazolium

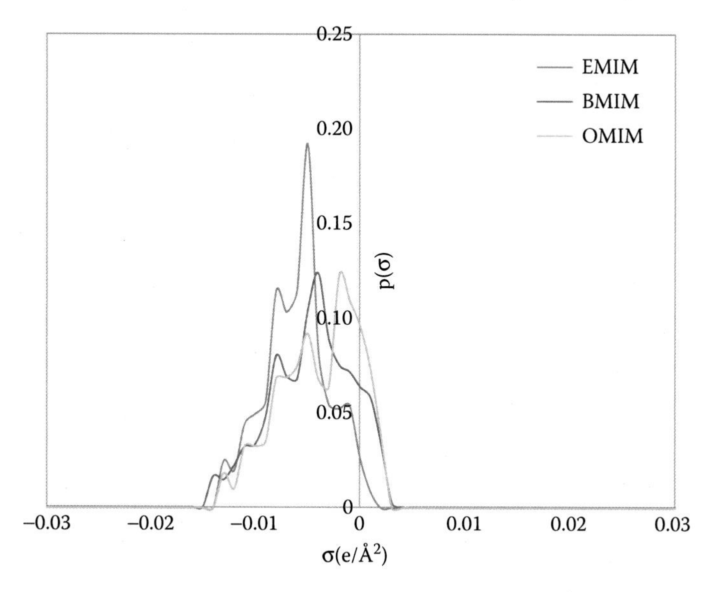

FIGURE 3.5
Sigma profiles of individual cations (EMIM: 1-ethyl-3-methylimidazolium; BMIM: 1-butyl-3-methylimidazolium; OMIM: 1-Octyl-3-methylimidazolium).

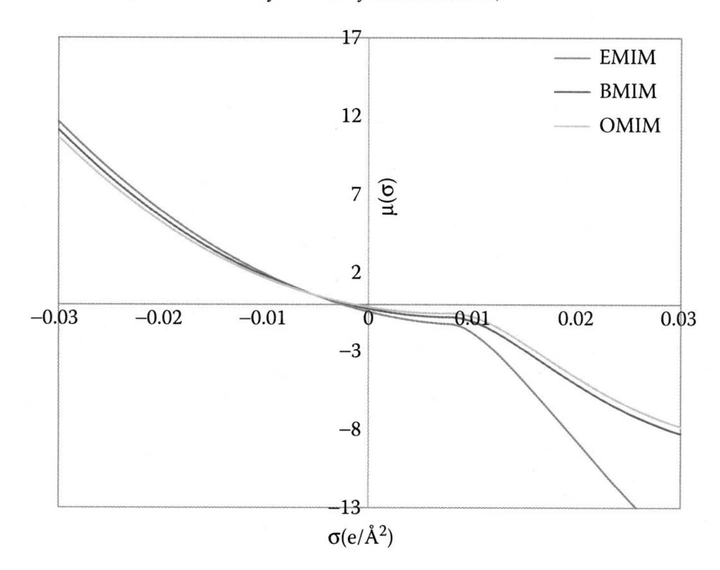

FIGURE 3.6
Sigma potential of individual cations.

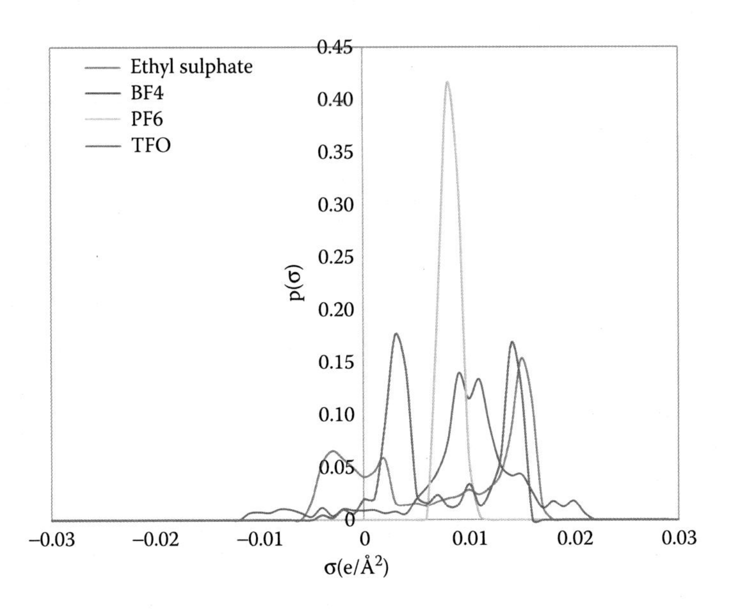

FIGURE 3.7
Sigma profile of individual anions.

([RMIM]; R=O, H, B etc.; O:Octyl; H:Hexyl; B:Butyl) (Figure 3.5) and then paired it up with three common anions, as given in Figure 3.7.

The sigma profiles and potentials for cations are shown in Figures 3.5 and 3.6, respectively. In all the normalized profiles, a small part of the sigma profile lies to the left of the hydrogen bonding cutoff radius (i.e. $\sigma < -0.0082$ e/Å^2), indicating a hydrogen bonding donor capacity. The sigma profiles for [BMIM] and [OMIM] are of the same nature. Most of the prominent peaks lie in the negative side of the sigma profile. The negative screening charges are due to the positive charge residing inside the imidazolium ring. [EMIM], [BMIM], [OMIM], all show peaks at the outmost position in the negative direction. They also show up in the sigma potential (Figure 3.6), that is, negative values of chemical potentials are encountered on the positive side of the screening charge densities (SCDs). The negative values are encountered since extra free energy is gained by forming hydrogen bonds. It should be noted that extra energy is gained with the formation of hydrogen bonds; this energy is negative in value when compared with the free energy required (which is positive in nature) for removing the SCDs.

The sigma profiles and potentials for anions are shown in Figures 3.7 and 3.8, respectively. For the anions, it is evident that a negative contribution to the sigma potential is due to positive screening charges, as obtained from their sigma profiles. It should be noted that the contribution of chemical or segment potential is a sum of two values, that is, the misfit term and the hydrogen bond term. For the anions, a negative value on the profile implies that it is the hydrogen-acceptor bonding that dominates the left-hand side of

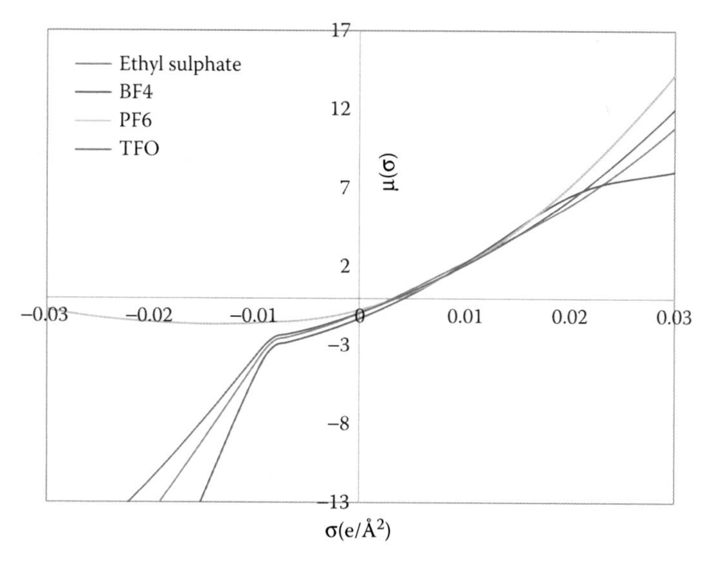

FIGURE 3.8
Sigma potentials of individual anions.

Figure 3.7. However, this is the opposite in the donor region, where the peak rises sharply owing to the difference in the shape and size of the segments or the misfit contribution is known to dominate the total interaction.

For all the anions, peaks are lying on the right of the cutoff zone for hydrogen bonding, that is, σ_{hb} = +0.0082 e/Å^2. All the prominent peaks for the anions, that is, 0.0085 e/Å^2 for [PF_6], 0.014 e/Å^2 for [CF_3SO_3], 0.015 e/Å^2 for [$EtSO_4$] and 0.011 e/Å^2 for [BF_4], lie to the right of the cutoff SCD, that is, σ_{hb} = +0.0082. This is due to the inherent negative charges of the anions. The sigma potential is a similar one as observed for cations, except that negative values of chemical potentials are encountered on the negative side of the SCDs.

3.2.8 Restoring Free Energy

Finally, we obtain the restoring free energy of molecule in solvent or mixture as

$$\frac{\Delta G_{i/s}^{*\text{res}}}{RT} = n_i \sum_{\sigma} p_i(\sigma_\alpha) \ln \Gamma_s(\sigma_\alpha) \tag{3.62}$$

This original definition was given by Klamt et al. (1998). Although this gave significant improvement for in hydrogen bonded systems, it was not fully consistent with the experimental observations. It should be noted that such policy made the hydrogen bonding contribution biased towards acceptors, which are limited to small electronegative atoms such as nitrogen, oxygen and fluorine. On the other hand, the donors originates from the hydrogen atoms which are in turn bonded to these electronegative atom(s). Hence, the process needs some course correction, as it leads to occasional meaningful errors. These weaknesses leave the process prone to occasional errors.

To incorporate a more meaningful hydrogen bond model, the definitions of acceptor and donor were modified. We assumed positively charged segments from the surfaces of all nitrogen, oxygen and fluorine atoms to be considered as acceptors. Similarly, the negatively charged segments from the surfaces of hydrogen atoms bound to any one of the acceptor atoms were chosen as HB donors. The modification will be taken up with test cases in Chapter 5 of this book. Now let us revert to Equation 3.7, which is given below:

$$
\begin{aligned}
&= \left(\frac{\Delta G_{i/S}^{*ES} - \Delta G_{i/i}^{*ES}}{kT} \right) + \left(\frac{\Delta A_{i/S}^{*vdW} - \Delta A_{i/i}^{*vdW}}{kT} \right) \\
&+ \left(\frac{\Delta A_{i/S}^{*\text{cav}} - \Delta A_{i/i}^{*\text{cav}}}{kT} \right) + \ln\left(\frac{V_{i/i}}{V_{i/S}} \right)
\end{aligned}
\tag{3.63}
$$

Here, we have obtained relation for the electrostatic contribution (first part of Equation 3.7, as above).

Now, we move our attention to the second term, that is, the dispersion or the van der Waals expression. This is the contribution due to the Helmholtz free function, which is modeled using a first order mean-free approach, the perturbation approach. For pure component, the expression takes the form:

$$
\frac{\Delta A_{i/i}^{*\text{disp}}}{NkT} = \frac{1}{kTV^{\text{mix}}} \sum_{i}^{r} \sum_{k}^{r} \varepsilon_{jk} m_j^i m_k^i \tag{3.64}
$$

Here, $\varepsilon_{jk} = \sqrt{\varepsilon_k \varepsilon_i}$ (K Å³) is the pair interaction energy between atom type j and atom type k, while m_j^i is the effective number of atoms of type j within a species i molecule, which is obtained from:

$$
m_j^i = \sum_{\alpha \in i} \left(\frac{S_a}{S_{a0}} \right)^{q_s} \tag{3.65}
$$

Where, S_a is the solvent accessible area of atom a and S_{a0} is the bare surface area calculated using the set of atomic radii in the quantum chemical calculations. q_s is the scaling factor in the range from 0 to unity. k is the Boltzmann constant in J/(molecule K), while V is the volume in Å³/molecule. Here, r denotes the atomic type: C, H, N, O, F, Cl, S, P, Br or I. For mixtures, it takes the following form (the interaction here is considered using all possible element pairs):

$$\frac{\Delta A_{\text{mix}}^{*\text{disp}}}{NkT} = -\frac{1}{kTV^{\text{mix}}} \sum_{\alpha=a} \sum_{\beta=a} \sum_{i\in\alpha}^{r} \sum_{j\in\beta}^{r} x_\alpha x_\beta m_i m_j \sqrt{\varepsilon_i \varepsilon_j} \quad (3.66)$$

Here, $\Delta A_{\text{mix}}^{*\text{disp}}$ is the dispersion free energy for the mixture in joules. N is the number of molecules, where $N = \Sigma_\alpha N_\alpha$ and V^{mix} the volume of the ideal mixture given by

$$V^{\text{mix}} = \sum_\alpha x_\alpha V_\alpha \quad (3.67)$$

x_α is the mole fraction given by: $x_\alpha = N_\alpha / N$. Here, i and j sum over all the components *bond types*, namely α and β. a refers to the number of components in the mixture. This contribution is used to account for all possible interactions between the atoms in different ensembles. For a binary mixture (Wang et al., 2007):

$$\frac{\Delta A_{\text{mix}}^{*\text{disp}}}{NkT} = -\frac{1}{TV^{\text{mix}}} \left[x_a^2 \sum_{i\in a}^{r} \sum_{j\in a}^{r} m_i m_j \sqrt{\varepsilon_i \varepsilon_j} + 2x_a x_b \sum_{i\in a}^{r} \sum_{j\in b}^{r} m_i m_j \sqrt{\varepsilon_i \varepsilon_j} + x_b^2 \sum_{i\in b}^{r} \sum_{j\in b}^{r} m_i m_j \sqrt{\varepsilon_i \varepsilon_j} \right] \quad (3.68)$$

This can be reduced as

$$\frac{\Delta A_{\text{mix}}^{*\text{disp}}}{NkT} = -\frac{1}{TV^{\text{mix}}} \left[x_a^2 E_{aa} + 2x_a x_b E_{ab} + x_b^2 E_{bb} \right] \quad (3.69)$$

Here

$$E_{aa} = \sum_{i\in a}^{r} \sum_{j\in a}^{r} m_i m_j \sqrt{\varepsilon_i \varepsilon_j} \qquad E_{ab} = \sum_{i\in a}^{r} \sum_{j\in b}^{r} m_i m_j \sqrt{\varepsilon_i \varepsilon_j} \quad (3.70)$$

The dispersion term of the residual chemical potential for species a and b is given as

$$\ln \gamma_i (\text{disp}) = \frac{\mu_i^{\text{disp}*}}{kT} \quad (3.71)$$

Hence, we obtain:

$$\frac{\mu_a^{\text{disp*}}}{kT} = \frac{1}{kT}\left(\frac{\partial \Delta A_{\text{mix}}^{*\text{disp}}}{\partial N_a}\right)_{T,V,N_{a\neq b}}$$
$$\Rightarrow -\frac{1}{T}\left[\frac{V^a}{V^{\text{mix}}}\left(x_a^2 E_{aa} + 2x_a x_b E_{ab} + x_b^2 E_{bb}\right) - 2\left(\frac{x_a E_{aa} + x_b E_{ab}}{V^{\text{mix}}}\right)\right] \tag{3.72}$$

Similarly, for bond type b, we have:

$$\frac{\mu_b^{\text{disp*}}}{kT} = \frac{1}{kT}\left(\frac{\partial \Delta A_{\text{mix}}^{*\text{disp}}}{\partial N_b}\right)_{T,V,N_{a\neq b}}$$
$$\Rightarrow -\frac{1}{T}\left[\frac{V^b}{V^{\text{mix}}}\left(x_a^2 E_{aa} + 2x_a x_b E_{ab} + x_b^2 E_{bb}\right) - 2\left(\frac{x_b E_{bb} + x_a E_{ab}}{V^{\text{mix}}}\right)\right] \tag{3.73}$$

For large complex molecules, additional corrections are required, which depend on the molecular size. It should be noted that the dispersion term is primarily responsible for the pure component calculations. However, it is the reverse for a mixture, where the contribution to the activity coefficient is negligible because of its cancellation between the solvent and pure solute phases. Hence, we shall neglect this contribution when we compute the activity coefficient.

Now again referring back to Equation 3.7, we will now concentrate on the remaining two terms, that is, $(\Delta A_{i/S}^{*\text{cav}} - \Delta A_{i/i}^{*\text{cav}}/kT)$ and $\ln(V_{i/i}/V_{i/S})$. These are calculated using the Staverman–Guggenheim contribution (Smith, Van Ness, & Abbott, 2001), where the liquid volumes can be accessed from the DIPPR database (Design Institute for Physical Properties, Sponsored by AIChE DIPPR Project 801). The contribution was derived based on a lattice model where the activity coefficient is caused by the difference in the shape and size of the molecule. It is given as

$$\ln\left(\gamma_i^{\text{comb}}\right) = 1 - \varphi_i + \ln(\varphi_i) - 5\frac{A_i^{\text{total}}}{A_{\text{norm}}}\left[1 - \frac{\varphi_i}{\theta_i} + \ln\left(\frac{\varphi_i}{\theta_i}\right)\right] \tag{3.74}$$

where:

$$\varphi_i = \frac{V_i^{\text{total}}}{\sum_j x_j V_j^{\text{total}}} \qquad \theta_i = \frac{A_i^{\text{total}}}{\sum_j x_j A_j^{\text{total}}} \tag{3.75}$$

Here, V_i^{total} is the volume of the molecule obtained from the COSMO file output (Figure 3.3). A_{norm} refers to a normalization constant, as A_i^{total} is not dimensionless. For molecules having similar shape and sizes, $\ln(\gamma_i^{\text{comb}}) \sim 0$, and hence, this term can be neglected. It should also be noted that $\ln(\gamma_i^{\text{comb}}) = 0$ for pure component, so it is used to model the cavity formation between a solvent and a solute only. We now are in a position to write the total expression for the activity coefficient of a component in the mixture. Thus, from Equations 3.62 through 3.65, we have the following equation:

$$\ln\left(\gamma_i^S x_i\right) = \left(\frac{\Delta G_{i/S}^{*ES} - \Delta G_{i/i}^{*ES}}{kT}\right) + \left(\frac{\Delta A_{i/S}^{*vdW} - \Delta A_{i/i}^{*vdW}}{kT}\right) + \left(\frac{\Delta A_{i/S}^{*\text{cav}} - \Delta A_{i/i}^{*\text{cav}}}{kT}\right) + \ln\left(\frac{V_{i/i}}{V_{i/S}}\right)$$

By writing out each contribution, we have

$$\frac{\Delta G_{i/s}^{*\text{res}}}{RT} = n_i \sum_{\sigma} p_i(\sigma_\alpha) \ln \Gamma_s(\sigma_\alpha)$$

$$\left(\frac{\Delta A_{i/S}^{*vdW} - \Delta A_{i/i}^{*vdW}}{kT}\right) = 0 \quad \text{(for mixtures)}$$

$$\left(\frac{\Delta A_{i/S}^{*\text{cav}} - \Delta A_{i/i}^{*\text{cav}}}{kT}\right) + \ln\left(\frac{V_{i/i}}{V_{i/S}}\right) = \ln\left(\gamma_i^{\text{comb}}\right) = 1 - \varphi_i + \ln(\varphi_i) - 5\frac{A_i^{\text{total}}}{A_{\text{norm}}}\left[1 - \frac{\varphi_i}{\theta_i} + \ln\left(\frac{\varphi_i}{\theta_i}\right)\right]$$

Thus, the activity coefficient of a segment that is $\Gamma_s(\sigma)$ and sigma potential are connected through the following relation:

$$p_s(\sigma_\alpha)(\Gamma_s(\sigma_\alpha)) = \exp\left(\frac{\mu_s(\sigma_\alpha)}{kT}\right) \tag{3.76}$$

The activity coefficient of segment in the mixture and in the pure liquid, $\Gamma_s(\sigma)$ and $\Gamma_i(\sigma)$, can then be rewritten as

$$\ln \Gamma_s(\sigma_\alpha) = -\ln\left\{\sum_{\sigma_n} p_s(\sigma_\alpha)\Gamma_s(\sigma_\alpha)\exp\left[-\frac{E^{\text{pair}}(\sigma_\alpha, \sigma_\beta)}{kT}\right]\right\} \tag{3.77}$$

The final expression then takes the form:

$$\ln \gamma_{i/S} = n_i \sum_{\sigma_m} p_i(\sigma_\alpha)\left[\ln \Gamma_s(\sigma_\alpha) - \ln \Gamma_i(\sigma_\alpha)\right] + \ln \gamma_{i/S}^{\text{comb}} \tag{3.78}$$

This equation is referred to as the COSMO Segment Activity Coefficient (COSMO-SAC) model. Now, we shall move ahead to compute the activity coefficient using COSMO-SAC model (Equation 3.78), as these values will be required to predict the tie line composition in case of ternary or binary system. The entire model is characterized by our earlier COSMO-SAC parameters, a_{eff} = 6.32 Å^2 (surface area of a standard segment), α' = 8419 kcal Å^4 mol^{-1} e^{-2} (misfit energy constant) for misfit energy interaction, c_{hb} = 75006 kcal Å^4 mol^{-1} e^{-2} (hydrogen bonding energy constant) and σ_{hb} = 0.0084 e Å^{-2} (hydrogen bonding cutoff).

3.3 Predictions of Tie Lines

The equilibrium for any ternary liquid–liquid system is defined by the equation:

$$\gamma_i^I x_i^I = \gamma_i^{II} x_i^{II} \quad \text{where } i = 1,2,3 \tag{3.79}$$

Here, our aim is to compute the γ_i, the activity coefficient of component *I*, as per Equation 3.78 from COSMO-SAC model. These values will be different for different phases (*I* and *II*). The corresponding mole fractions in phases *I* and *II* will be denoted by x_i^I and x_i^{II}, respectively. Assuming an adiabatic operation and that both feed *F* and solvent *S* enter a theoretical stage at same temperature, we are then left to account for the heat of mixing. This value for most thermodynamics calculations involving liquid–liquid equilibria (LLE) is sufficiently small, as we observe that a small temperature change occurs. For calculating the LLE split, we adopt the modified Rachford–Rice algorithm (Figure 3.9). Initially, a random feed composition to any known LLE data is used. In any case, we can use the reported data to get an initial feed point. Thus, the feed concentration (z_i) is then calculated using the following equation:

$$z_i = \frac{x_i^I + x_i^{II}}{2} \tag{3.80}$$

Assuming a unity feed rate, that is, $F = 1$, we visualize the process of one feed flow, namely *F*, entering a flash column and then flashing into one liquid phase (*L*) and one vapor phase (*V*), such that $F = L + V$. In the next step, the values of distribution coefficient (K_i, I = 1,2,3) are then assigned as follows:

$$K_i = \frac{x_i^{II}}{x_i^I} = \frac{\gamma_i^I}{\gamma_i^{II}} \tag{3.81}$$

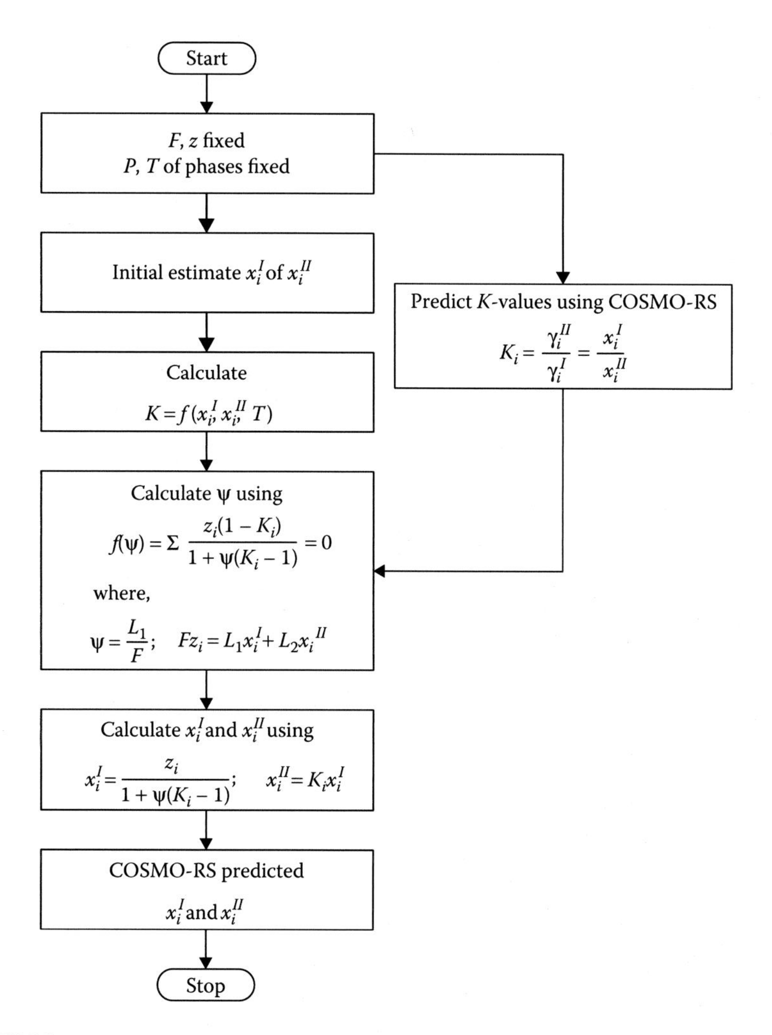

FIGURE 3.9
Modified Rachford–Rice algorithm for COSMO-SAC model.

Where, γ_i^I and γ_i^{II} are predicted using the COSMO-SAC model. With the values of K_i, the isothermal flash equation is solved using the conventional flash expression:

$$f(\Psi) = \sum \frac{z_i(1-K_i)}{1+\Psi(K_i-1)} = 0 \tag{3.82}$$

subject to,

$$Fz_i = L_1 x_i^I + L_2 x_i^{II} \text{ and } \Psi = L_1/F \tag{3.83}$$

Here, we have assumed $L = L_1$ and $V = L_2$, as it contains both liquid phases. Both the liquid phases represent the flow rate of the extract L_1 and the raffinate phase L_2. Equation 3.83 is nonlinear in nature. This is first iteratively solved for Ψ, which is a fraction and varies between zero and unity. Thereafter, the mole fractions in both phases are calculated via Equations 3.84 and 3.85, respectively.

$$x_i^I = \frac{z_i}{1+\Psi(K_i-1)} \tag{3.84}$$

$$x_i^{II} = K_i x_i^I \tag{3.85}$$

For the ionic liquid systems described in the book, the in-house experimental data were used to assume the feed compositions (Equation 3.80). Thus, in a summary, the COSMO-SAC model can predict the tie lines based on the initial input of F, z_i, P, T. Here, the effect of pressure on LLE computations has been assumed to be negligible. Thus, not only ionic liquids but also any unknown system can be predicted once its COSMO files (Figure 3.3) are generated. This also makes use of the fact that a successful computation of Ψ by Equation 3.82 indicates a heterogeneous region. A feed point outside the bimodal cure, as observed in a ternary LLE diagram, will usually fail to converge Equation 3.82. Thus, based on any random feed composition (z_i), one can generate a tie line data. These predictions can be further used for other higher-order, namely quaternary and quinary, systems. It has been described in detail for the computation of VLLE (vapor–liquid–liquid equilibria) diagrams by making necessary adjustments. This has been discussed in Chapter 4.

3.4 Predictions of LLE of Ionic Liquid Systems

Recent experimental work suggests that ILs can be considered as fully dissociated cations and anions. Further, the mechanism of hydrogen bonding (Scheiner, 1997) in the ILs is also not fully available. This fact was used to study the coordination chemistry extraction mechanisms of metal ions, particularly actinide (Cocalia, Gutowski, & Rogers, 2006; Jensen, Neuefeind, Beitz, Skanthakumar, & Soderholm, 2003; MacFarlane & Seddon, 2007). Experimentally, these prove that ILs are completely dissociated into cations and anions when in solution. Also, the definition of IL states it to be a liquid with melting point below 100°C and containing ions exhibiting ionic conductivity. The fact that the ionic conductivity can be measured experimentally proves that ILs consist of ions in solutions. Ionic conductivity studies have been carried out by Carda-Broch et al., (2003), Berthod and Armstrong (2003) on fluorinated anions such as $[PF_6]$.

Further, in molecular dynamic simulations, a cubical box is taken, which contains equal number of cations and anions, instead the whole IL molecule. For example, in the work of Morrow and Maginn (2002), the authors studied the thermo-physical properties of ionic liquids by using 300 cations and 300 anions. Similarly, in the work of Lee, Jung and Han (2005), the ionic conductivity of ILs were found using a cubical box of 100 cations and 100 anions. On the contrary, Earle et al. (2006) recently distilled IL without decomposition, at very low pressures. In their pioneering work, they were successful in distilling the $[(CF_3SO_2)_2N]$-based ILs at 100 Pa and 573 K, without decomposition. However, in LLE experiments, ILs are assumed to be in isobaric condition of 1 atm, whereas the distillation is observed with pressures ~0.000009 atm. Since the operating pressures of LLE are 1 atm, our approach holds true.

Dissociation was also observed during vapor–liquid equilibria (VLE) predictions on reported systems (Kato & Gmehling, 2005; Kato, Krummen, & Gmehling, 2004) in ILs, where we obtained better results than obtained using a single composite molecule (Diedenhofen, Eckert, & Klamt, 2003; Table 3.1). This prompted us to propose that the IL be considered as a dissociated pair of cation and anion for both the VLE and LLE predictions. This novel approach can be applied to ILs containing a limitless combination of

TABLE 3.1

VLE Comparison of RMSD with Different Models

			COSMO-RS	
No.	**Reference**	**System**	**Single Molecule**	**Cation and Anion**
1	(Döker & Gmehling, 2005)	$[Emim][(CF_3SO_2)_2N]$ + Acetone	2.70	1.81
2	(Döker & Gmehling, 2005)	$[Emim][(CF_3SO_2)_2N]$ + 2-Propanol	3.80	2.60
3	(Döker & Gmehling, 2005)	$[Emim][(CF_3SO_2)_2N]$ + Water	9.80	5.26
4	(Döker & Gmehling, 2005)	$[Bmim][(CF_3SO_2)_2N]$ + Acetone	3.50	3.08
5	(Döker & Gmehling, 2005)	$[Bmim][(CF_3SO_2)_2N]$ + 2-Propanol	4.00	2.89
6	(Döker & Gmehling, 2005)	$[Bmim][(CF_3SO_2)_2N]$ + Water	9.80	4.56
7	(Kato & Gmehling, 2005)	$[Mmim][(CH_3)_2PO_4]$ + Acetone	5.10	2.50
8	(Kato & Gmehling, 2005)	$[Mmim][(CH_3)_2PO_4]$ + Tetrahydrofuran	4.60	2.28
9	(Kato & Gmehling, 2005)	$[Mmim][(CH_3)_2PO_4]$ + Water	10.60	6.84
10	(Kato & Gmehling, 2005)	$[Mmim][(CF_3SO_2)_2N]$ + Benzene	8.92	3.66
11	(Kato & Gmehling, 2005)	$[Mmim][(CF_3SO_2)_2N]$ + Cyclohexane	6.26	4.23
12	(Kato et al., 2004)	$[Emim][EtSO_4]$ + Benzene	9.75	5.62
13	(Kato et al., 2004)	$[Emim][EtSO_4]$ + Cyclohexane	6.94	2.36
		Average	6.59	3.66

cation and anion (~10^{18}). Therefore, one has to do separate quantum mechanical calculations for cations and anions. Thereafter, one can test an IL for a particular application by addition of sigma profiles and sigma potential of cation and anion only.

Moreover, till date, an application concerned with the assumption of complete dissociation of ILs into cations and anions with equimolar concentrations has been carried out earlier by Klamt and Schuurmann (1993) has assumed a complete dissociation of cations and anions for their prediction of infinite dilution activity coefficients (IDAC) of solutes in ILs by COSMO-RS implementation. In their work, the ILs have been described by an equimolar mixture of two distinct ions, the cation and the anion, which finally contribute to the sigma profile of the mixture as two different compounds.

Hence, for the prediction of LLE-based system involving ILs, two modes of approach were assumed: (1) IL as a pure solvent and (2) IL consisting of a pair of cations and anions (Table 3.2). The latter approach assumes a solvent of a mixture of two components: cation and anion. The latter approach has proved to give better phase split with lower root-mean-square deviation. Hence, we proposed a complete dissociation of ILs into cations and anions with equimolar quantities, which is in line with the linear addition of the sigma profiles of the cation and anion.

$$p_{\text{ionic liquid}}(\sigma) = p_{\text{cation}}(\sigma) + p_{\text{anion}}(\sigma) \tag{3.86}$$

Where, $p_{\text{cation}}(\sigma)$ and $p_{\text{anion}}(\sigma)$ are the sigma profiles for cation and anion, respectively. This is based on the assumption that the IL in a solution is fully dissociated into its respective cations and anions with equimolar concentrations. It should be noted that the entropy of mixing will not exist, since we are dealing with a sigma profile of a single compound/mixture. Thereafter, the linear additions of COSMO area and volume are done, so as to obtain the profile of single mixture/compound.

The results of all the reported IL ternary systems are given in Table 3.2. Using the sigma profile for the single composite molecule, it is observed that for nearly half of the systems, the values of the activity coefficients were not able to predict the split between the extract (IL phase) and raffinate phase for the given tie line. Predictions giving an average RMSD of 36.5% and a maximum root-mean-square deviation (RMSD) of 90% ([Omim][Cl]-Benzene-Heptane) were observed. NO **SPLIT (NS)** condition for 15 systems was encountered. For 13 systems, an RMSD much greater than 10% was observed.

The data sets contain 12 different ILs comprising seven different cations and eight anions. Apart from imidazolium-based ILs, pyridinium-based IL (1-butyl-3-methylpyridinium tetrafluoroborate [Bmpy][BF_4]) has also been studied. Switching from composite to additive sigma profile, we notice that a 'NO SPLIT' situation has been converted to a 'SPLIT' between the two phases for all the 15 cases. It is clear that the improvement is drastic for the IL containing system namely [Bmim][CF_3SO_3], [Omim][MDeg] and [Bmpy][BF_4]

TABLE 3.2

LLE Comparison of Single and Cation–Anion Pair by Using COSMO-RS

			Root-Mean-Square Deviation		Experiment
System No.	**T/K**	**System**	**Single Molecule**	**Cation + Anion**	**Error**
1	298.15	[Bmim][(CF_3SO_3] – Ethanol-ethyl-*tert*-butyl ether	NS	0.095	NA
2	298.15	[Bmim][(CF_3SO_3] – Ethanol-*tert*-amyl ethyl ether	NS	0.032	0.004
3	298.15	[Hmim][BF_4] – Benzene-heptane	0.711	0.043	0.004
4	298.15	[Hmim][BF_4] – Benzene-dodecane	0.087	0.098	0.004
5	298.15	[Hmim][BF_4] – Benzene-hexadecane	0.481	0.072	0.004
6	298.15	[Hmim][BF_4] – Ethanol-hexene	NS	0.147	0.004
7	298.15	[Hmim][BF_4] – Ethanol-heptene	NS	0.125	0.004
8	298.15	[Hmim][PF_6] – Benzene-heptane	0.856	0.042	0.004
9	298.15	[Hmim][PF_6] – Benzene-dodecane	0.385	0.047	0.004
10	298.15	[Hmim][PF_6] – Benzene-hexadecane	0.457	0.059	0.004
11	298.15	[Hmim][PF_6] – Ethanol-hexene	NS	0.315	0.004
12	298.15	[Hmim][PF_6] – Ethanol-heptene	NS	0.306	0.004
13	298.15	[Omim][Cl] – Methanol-hexadecane	0.029	0.005	NA
14	298.15	[Omim][Cl] – Ethanol-hexadecane	0.03	0.008	NA
15	298.15	[Omim][Cl] – Ethanol-*tert*-amyl ethyl ether	0.382	0.067	NA
16	298.15	[Omim][Cl] – Benzene-heptane	0.896	0.096	0.006
17	298.15	[Omim][Cl] – Benzene-dodecane	0.058	0.096	0.006
18	298.15	[Omim][Cl] – Benzene-hexadecane	0.146	0.07	0.006
19	298.15	[Emim][$C_8H_{17}SO_4$] – Benzene-heptane	0.362	0.07	0.006
20	298.15	[Emim][$C_8H_{17}SO_4$] – Benzene-hexadecane	0.456	0.14	0.006

(*Continued*)

TABLE 3.2 (*Continued*)

LLE Comparison of Single and Cation–Anion Pair by Using COSMO-RS

			Root-Mean-Square Deviation		Experiment
System No.	T/K	System	Single Molecule	Cation + Anion	Error
21	298.15	[Omim][MDEG][a] – Benzene-heptane	NS	0.103	0.006
22	298.15	[Omim][MDEG][a] – Benzene-hexadecane	NS	0.043	0.006
23	313.15	[Bmpy][BF_4] – Xylene-octane	NS	0.03	0.0025
24	348.15	[Bmpy][BF_4] – Xylene-Octane	NS	0.024	0.0025
25	313.15	[Bmpy][BF_4] – Ethylbenzene-octane	NS	0.041	0.0025
26	348.15	[Bmpy][BF_4] – Ethylbenzene-octane	NS	0.029	0.0025
27	313.15	[Bmpy][BF_4] – Benzene-hexane	NS	0.048	0.0025
28	333.15	[Bmpy][BF_4] – Benzene-hexane	NS	0.036	0.0025
29	298.15	[Bmpy][BF_4] – Toluene-heptane	NS	0.075	0.0025
30	298.15	[Mmim][CH_3SO_4] – Toluene-heptane	0.455	0.008	0.0025
31	298.15	[Emim][$C_2H_5SO_4$] – Toluene-heptane	0.116	0.031	0.0025
32	298.15	[Bmim][[CH_3SO_4] – Toluene-heptane	0.235	0.032	0.0025
33	313.15	[Emim][$EtSO_4$] – Ethanol-hexene	NS	0.124	0.005
34	313.15	[Emim][$EtSO_4$] – Ethanol-heptene	NS	0.023	0.005
35	313.15	[E-2,3-dmim][$EtSO_4$] – Ethanol-hexene	NS	0.032	0.005
36	313.15	[E-2,3-dmim][$EtSO_4$] – Ethanol-heptene	NS	0.065	0.005

Note: NS: No splitting of the phase; NA: Not available

[a] MDEG: monomethyldiethyleneglycol

systems and also for the alkene-based systems [System No. 6, 7, 11, 12], for which no prediction had been possible when considering the IL as a single molecule. For the [Bmim][CF_3SO_3], [Omim][MDeg] and [Bmpy][BF_4] systems (System No. 1, 2, 21, 22 and 23–29), the RMSD is less than 10%. Out of the 36 systems studied, most of the predictions are excellent, considering the fact that only six systems gave RMSD greater than 10%.

Acknowledgments

A part of this chapter has been adapted from my earlier National Program for Technology Enhanced Learning (NPTEL) web course entitled 'Molecular Simulation in Chemical Engineering'. I am deeply indebted to the Ministry of Human Resource and Development, Government of India, for providing me financial support for the same. Further, I would like to thank Prof. Stanley I. Sandler, University of Delaware, for helping me in learning the basics of COSMO-based modeling. I am also indebted to his student Dr. Russel I. Burnett for helping me out in the COSMO-SAC equations. This visit would not have been possible without the funding from IUSSTF (Indo–U.S. Science and Technology Forum).

Appendix: Canonical Partition Functions

The canonical partition function that is $Q(N, V, \beta)$ is given as

$$Q(N,V,\beta) = \sum_{\substack{\text{states} \\ i}} e^{-\beta E_i(N,V)} \tag{A.1}$$

The probability of occurrence of a particular miscrostate i with Energy E_α is given as

$$p_i(E_\alpha) = \frac{e^{-\beta E_\alpha}}{\sum_{\text{All } i \text{ states}} \mathrm{e}^{-\beta E_i}} = \frac{e^{-\beta E_\alpha}}{Q(N,V,\beta)} \tag{A.2}$$

The significance of β can be made from the fact that as $\beta > 0$, the state of higher energy is less likely as the state of lower energy. We now shall see the process of evaluating the thermodynamic properties based on canonical partition functions. It should be noted that the probability of finding a system in any microstate of energy E_α would be the product of the probability that the system is in a particular microstate with energy E_α and the number of states having that energy (also known as degeneracy). Thus, we have:

$$p(E_\alpha) = \sum_{\substack{\text{States } i \text{ having} \\ \text{Energy } E_\alpha}} p_i(E_\alpha) = p_i(E_\alpha) \times \text{Degeneracy}\begin{pmatrix}\text{number of states} \\ \text{with energy } E_\varepsilon\end{pmatrix} \tag{A.3}$$

We now define the probability of occurrence of a particular microstate i with energy E_α as

$$p_i(E_\alpha) = \frac{e^{-\beta E_\alpha}}{\sum_{\text{All } i \text{ states}} e^{-\beta E_i}} = \frac{e^{-\beta E_\alpha}}{Q(N,V,\beta)} \tag{A.4}$$

In a similar manner, the probability of occurrence of the energy level E_α, $p(E_\alpha)$ is

$$p(E_\alpha) = \text{Degeneracy}\begin{pmatrix}\text{number of states} \\ \text{with energy } E_\varepsilon\end{pmatrix} \times p_i(E_\alpha) = \frac{\omega(E_\alpha)e^{-\beta E_\alpha}}{Q(N,V,\beta)} \tag{A.5}$$

Here Degeneracy = $\omega(E_\alpha)$

Thus, the partition function can be written in terms of either levels or states, as below:

$$Q(N,V,\beta) = \sum_{\text{All } i \text{ states}} e^{-\beta E_i} = \sum_{\text{All } j \text{ levels}} \omega(E_j)e^{-\beta E_i} p_i(E_\alpha) \tag{A.6}$$

Now, once the partition function has been defined, we can derive thermodynamics functions such as internal energy U:

$$U = \sum_{\text{all states of } k} E_k p(E_k) = \frac{\sum_{\text{all states of } k} E_k p(E_k)}{\sum_{\text{all states of } k} e^{-\beta E_k}} = \frac{\sum_{\text{all states of } k} E_k p(E_k)}{Q} \tag{A.7}$$

However, from the definition, we have:

$$\left(\frac{\partial Q}{\partial \beta}\right)_{N,V} = -\sum_{k \text{ states}} E_k e^{-\beta E_k} \tag{A.8}$$

Thus, the internal energy U takes the form:

$$U = \left(\frac{\partial \ln Q}{\partial \beta}\right)_{N,V} = \frac{\sum_{k \text{ states}} E_k e^{-\beta E_k}}{Q} \tag{A.9}$$

References

Banerjee, T., Singh, M. K., & Khanna, A. (2006). Prediction of binary VLE for imidazolium based ionic liquid systems using COSMO-RS. *Industrial & Engineering Chemistry Research, 45*(9), 3207–3219. doi:10.1021/ie051116c.

Burnett, R. I. (2012). *Predicting liquid-phase thermodynamic properties using COSMO-SAC*. PhD Thesis, University of Delaware, Newark, DE.

Carda-Broch, S., Berthod, A., & Armstrong, D. W. (2003). Solvent properties of the 1-butyl-3-methylimidazolium hexafluorophosphate ionic liquid. *Analytical and Bioanalytical Chemistry, 375*(2), 191–199. doi:10.1007/s00216-002-1684-1.

Cocalia, V. A., Gutowski, K. E., & Rogers, R. D. (2006). The coordination chemistry of actinides in ionic liquids: A review of experiment and simulation. *Coordination Chemistry Reviews, 250*(7–8), 755–764. doi:10.1016/j.ccr.2005.09.019.

Design Institute for Physical Properties, Sponsored by AIChE DIPPR Project 801 – Full Version. Design Institute for Physical Property Data/AIChE.

Diedenhofen, M., Eckert, F., & Klamt, A. (2003). Prediction of infinite dilution activity coefficients of organic compounds in ionic liquids using COSMO-RS. *Journal of Chemical & Engineering Data, 48*(3), 475–479. doi:10.1021/je025626e.

Döker, M., & Gmehling, J. (2005). Measurement and prediction of vapor–liquid equilibria of ternary systems containing ionic liquids. *Fluid Phase Equilibria, 227*(2), 255–266. doi:10.1016/j.fluid.2004.11.010.

Earle, M. J., Esperanca, J. M. S. S., Gilea, M. A., Canongia Lopes, J. N., Rebelo, L. P. N., Magee, J. W., … Widegren, J. A. (2006). The distillation and volatility of ionic liquids. *Nature, 439*(7078), 831–834. doi:10.1038/nature04451.

Grensemann, H., & Gmehling, J. (2005). Performance of a conductor-like screening model for real solvents model in comparison to classical group contribution methods. *Industrial & Engineering Chemistry Research, 44*(5), 1610–1624. doi:10.1021/ie049139z.

Jensen, M. P., Neuefeind, J., Beitz, J. V., Skanthakumar, S., & Soderholm, L. (2003). Mechanisms of metal ion transfer into room-temperature ionic liquids: The role of anion exchange. *Journal of the American Chemical Society, 125*(50), 15466–15473. doi:10.1021/ja037577b.

Kato, R., & Gmehling, J. (2005). Measurement and correlation of vapor–liquid equilibria of binary systems containing the ionic liquids [EMIM][$(CF_3SO_2)_2N$], [BMIM][$(CF_3SO_2)_2N$], [MMIM][$(CH_3)_2PO_4$] and oxygenated organic compounds respectively water. *Fluid Phase Equilibria, 231*(1), 38–43. doi:10.1016/j.fluid.2005.01.002.

Kato, R., Krummen, M., & Gmehling, J. (2004). Measurement and correlation of vapor–liquid equilibria and excess enthalpies of binary systems containing ionic liquids and hydrocarbons. *Fluid Phase Equilibria, 224*(1), 47–54. doi:10.1016/j.fluid.2004.05.009.

Klamt, A. (1995). Conductor-like screening model for real solvents: A new approach to the quantitative calculation of solvation phenomena. *The Journal of Physical Chemistry, 99*(7), 2224–2235. doi:10.1021/j100007a062.

Klamt, A., Jonas, V., Bürger, T., & Lohrenz, J. C. W. (1998). Refinement and parametrization of COSMO-RS. *The Journal of Physical Chemistry A, 102*(26), 5074–5085. doi:10.1021/jp980017s.

Klamt, A., & Schuurmann, G. (1993). COSMO: A new approach to dielectric screening in solvents with explicit expressions for the screening energy and its gradient. *Journal of the Chemical Society, Perkin Transactions* 2(5), 799–805. doi:10.1039/P29930000799.

Klamt, A. (2005). *COSMO-RS: From quantum chemistry to fluid phase thermodynamics and drug design*. Amsterdam, the Netherlands: Elsevier.

Lee, S. U., Jung, J., & Han, Y.-K. (2005). Molecular dynamics study of the ionic conductivity of 1-n-butyl-3-methylimidazolium salts as ionic liquids. *Chemical Physics Letters, 406*(4–6), 332–340. doi:10.1016/j.cplett.2005.02.109.

Lin, S.-T., Chang, J., Wang, S., Goddard, W. A., & Sandler, S. I. (2004). Prediction of vapor pressures and enthalpies of vaporization using a COSMO solvation model. *The Journal of Physical Chemistry A, 108*(36), 7429–7439. doi:10.1021/jp048813n.

Lin, S.-T., & Sandler, S. I. (2002). A priori phase equilibrium prediction from a segment contribution solvation model. *Industrial & Engineering Chemistry Research, 41*(5), 899–913. doi:10.1021/ie001047w.

MacFarlane, D. R., & Seddon, K. R. (2007). Ionic liquids–Progress on the fundamental issues. *Australian Journal of Chemistry, 60*(1), 3–5. doi:10.1071/CH06478.

Morrow, T. I., & Maginn, E. J. (2002). Molecular dynamics study of the ionic liquid 1-n-Butyl-3-methylimidazolium hexafluorophosphate. *The Journal of Physical Chemistry B, 106*(49), 12807–12813. doi:10.1021/jp0267003.

Pye, C. C., Ziegler, T., van Lenthe, E., & Louwen, J. N. (2009). An implementation of the conductor-like screening model of solvation within the Amsterdam density functional package–Part II. COSMO for real solvents. *Canadian Journal of Chemistry, 87*(7), 790–797. doi:10.1139/V09-008.

Scheiner, S. (1997). *Hydrogen bonding: A theoretical perspective*. New York: Oxford University Press.

Smith, J. M., Van Ness, H. C., Abbott, M. M. (2001). *Introduction to chemical engineering Thermodynamics* (6th ed.). Boston, MA: McGraw-Hill.

Wang, S., Lin, S.-T., Chang, J., Goddard, W. A., & Sandler, S. I. (2006). Application of the COSMO–SAC–BP solvation model to predictions of normal boiling temperatures for environmentally significant substances. *Industrial & Engineering Chemistry Research, 45*(16), 5426–5434. doi:10.1021/ie050352k.

Wang, S., Sandler, S. I., & Chen, C.-C. (2007). Refinement of COSMO–SAC and the Applications. *Industrial & Engineering Chemistry Research, 46*(22), 7275–7288. doi:10.1021/ie070465z.Wang, S., Stubbs, J. M., Siepmann, J. I., & Sandler, S. I. (2005). Effects of conformational distributions on sigma profiles in COSMO theories. *The Journal of Physical Chemistry A, 109*(49), 11285–11294. doi:10.1021/jp053859h.

4

Application of COSMO-SAC in Complex Phase Behavior: Vapor–Liquid–Liquid Equilibria

4.1 Introduction

An integrated part of chemical engineering is the calculation of phase equilibria. The importance of phase equilibria calculations is realized in designing the process equipment and optimizing separation and purification processes. Phase equilibria involves either homogeneous phases, that is, two liquid phases or heterogeneous phases, for example, liquid with vapor or solid phases. The main objective of these calculations is to recognize the presence of one vapor or liquid, two vapor–liquid or liquid–liquid, or multiple phases. Thus, the knowledge of thermodynamic properties of pure and mixture fluids is absolutely essential to identify these regions. For decades, researchers and process engineers established phase equilibria calculations both experimentally and computationally. Though experimental database are well established for binary and ternary vapor–liquid equilibria (VLE) and liquid–liquid equilibria (LLE), they are time consuming and expensive. Moreover, database is not so strong for complex-phase equilibria processes. The lack of experimental data and the search for a priori prediction motivate researchers to develop thermodynamic models for various chemical and pharmaceutical processes. The modern-day process simulators such as ChemCAD and ASPEN run these models to get a priori phase equilibria. In Chapter 2, we have discussed computation techniques of LLE with three state-of-the-art industrial problems. This chapter aims to provide a detailed analysis of complex-phase behavior involving two liquid phases and one vapor phase simultaneously present, better known as the vapor–liquid–liquid equilibria (VLLE).

Information regarding temperature, pressure and composition in VLE and LLE is crucial for designing distillation and extraction processes. However, thermodynamic properties are necessary in complex-phase behaviors like VLLE. If the presence of heterogeneous liquid mixtures is not correctly accounted for the systems that exhibit a miscibility gap, there may be multiple solutions to vapor–liquid phases, which in turn are a possible reason for multiple steady states in heterogeneous distillation. VLLE is frequently encountered in heterogeneous azeotropic distillation which is widely found in hydrocarbon industry. The heterogenic azeotropic distillation separates azeotropic mixtures into their components. The thermodynamic models, used to compute heterogeneous azeotropic distillation, should describe VLE and LLE as accurately as possible (Wyczesany, 2014). In this chapter, we will highlight the experimental and computational techniques used for VLLE. A priori based COSMO-SAC model (discussed in Chapter 3) will be used to predict complex-phase behaviors of VLLE and the calculation procedure will be elaborated with eight state-of-the-art industrial problems.

4.2 Thermodynamics of VLLE

The criteria for phase equilibria in a closed, nonreacting multi-component system at constant energy and volume is that the pressure must be same on both sides and the partial molar Gibbs free energy of each species must be same in each phase (Sandler, 1999). The criterion that forms basis of phase equilibrium calculation is expressed mathematically by Equation 4.1.

$$\overline{G}_i^I = \overline{G}_i^{II} \tag{4.1}$$

For the VLLE, the vapor phase is in equilibrium with individual liquid phases and liquid phases are in equilibrium between them too. Therefore,

$$\overline{G}_i^{L_I} = \overline{G}_i^{L_{II}} = \overline{G}_i^{V} \tag{4.2}$$

(liquid L_I, L_{II} and vapor V)

In terms of fugacity, we can write,

$$\overline{f}_i^{L_I} = \overline{f}_i^{L_{II}} = \overline{f}_i^{V} \tag{4.3}$$

We explain the fugacity term by the activity coefficient model. By neglecting fugacity coefficient corrections and considering total pressure as well as vapor pressure of the species are sufficiently low,

$$y_i P = x_i \gamma_i P_i^{\text{vap}} \tag{4.4}$$

Further, from LLE, we get,

$$x_i^I \gamma_i^I = x_i^{II} \gamma_i^{II} \tag{4.5}$$

Thus, combining Equations 4.4 and 4.5, we get the equilibrium relationship for VLLE.

$$y_i P = x_i^I \gamma_i^I P_i^{\text{vap}} = x_i^{II} \gamma_i^{II} P_i^{\text{vap}} \quad (4.6)$$

$$\frac{x_i^I}{x_i^{II}} = \frac{\gamma_i^{II}}{\gamma_i^I} = K_i \quad (4.7)$$

where K_i is known as the distribution coefficient and is the ratio of solute concentration in two phases. When solutes are added into two partially miscible or immiscible solvents, these unequally distribute in the two liquid phases. The distribution coefficient can efficiently be calculated from excess Gibbs free energy models which will be discussed in Section 4.4.

4.3 Experimental Procedure of VLLE

There are five main procedures, broadly classified into isobaric and isothermal procedures, by which VLLE experiment can be carried out. The isobaric procedure is divided further into distillation, dynamic and flow methods, and isothermal procedures consist of dew- and bubble-point methods. The experimental setup for VLLE mainly uses modified VLE setup, and therefore, eases to get a heterogeneous liquid region. A brief description of the isobaric method for obtaining VLLE is given in this section. The first method to obtain VLLE data is the distillation method. To make the liquid phase heterogeneous, a stirrer was introduced inside the equipment. A small portion of liquid from boiling flask is allowed to distil. This boiling flask is part of a liquid mixture of known composition. Due to a large quantity of liquid mixture, the distillation process does not change compositions. The distillate is later analyzed to determine the composition of vapor in equilibrium. The most used method to date to measure VLLE is the dynamic method. Based on the Gillespie principle, the method works on the heterogeneous liquid region. An ultrasonic homogenizer in a distilling flask was introduced by Gomis, Ruiz, and Asensi (2000) to emulsify the liquid phases and thus for rapid mass transfer between the two phases. The dynamic method is difficult to attain equilibrium for heterogeneous liquid mixtures at the boiling temperature. The flow method attains steady state in little time because the equilibrium chamber is constantly fed with constant composition steady-state feed stream. A wide variety of equipment designs and operating principles are available for this method.

4.4 Computational Techniques of VLLE

In Chapter 2, we introduced excess Gibbs free energy based models such as nonrandom two-liquid (NRTL) and UNIversal QUAsiChemical (UNIQUAC). These two models correlate VLE and LLE with high precision. Nevertheless, coefficients describing the VLE exactly predict the LLE very inaccurately and vice versa (Wyczesany, 2014). In VLLE, vapor–liquid and liquid–liquid coexisting regions overlap making it further challenging to estimate the parameters. The computations are divided into two broad classes of techniques, namely the equation-solving approach and the Gibbs free energy minimization approach. The equation-solving techniques combine mass balances and equilibrium relations into a set of nonlinear algebraic equations, while the Gibbs energy minimization approach formulates the phase equilibrium problem into an optimization problem which can be solved by a global solver subject to material balance constraints (Alsaifi & Englezos, 2011). The excess Gibbs free energy based models deal with the nonideality of the system. Examples for which are NRTL (Renon & Prausnitz, 1968), UNIQUAC (Abrams & Prausnitz, 1975), UNIFAC (Fredenslund, Jones, & Prausnitz, 1975) and COSMO-SAC (Lin & Sandler, 2002). The first three models are heavily dependent on the experimental data set. The binary interaction parameters are optimized over a vast data range of experimental data and fed into process simulators. The dependency of an experimental data set is less for the UNIFAC model as it considers interactions among functional groups.

Models based on an equation of state (EoS) are quite popular for thermodynamic modelling. Since Van der Waals first modified the ideal gas law and introduced the two-parameter EoS, several others evolved to predict thermodynamic properties. Most notable of which are Peng–Robinson (PR; Peng & Robinson, 1976) and Soave–Redlich–Kwong (SRK; Soave, 1972). The expression for PR is

$$P = \frac{RT}{\underline{v} - b} - \frac{a}{\underline{v}(\underline{v} + b) + b(\underline{v} - b)} \tag{4.8}$$

where:
 a is the temperature-dependent energy parameter
 b is the volume parameter.

Traditionally, they are calculated from critical properties and an accentric factor of a chemical substance. An alternative approach of calculating a and b from first principle solvation charging free energy is proposed by Hsieh and Lin (2009a, 2009b, 2010) in a series of papers. Later, they applied it in VLLE by fusing concepts of PR and COSMO-SAC. The two parameters are given by

$$a(T, x) = \frac{b}{C_{PR}} \Delta G^{*\text{chg}}(T, x) \tag{4.9}$$

$$b(x)=\sum x_i b_i \tag{4.10}$$

where x_i is the mole fraction of substance i in mixture and b_i is the volume parameter of species i and can be obtained from COSMO calculation. $\Delta G^{*chg}(T,x)$ is the total solvation charging free energy and is obtained from

$$\Delta G^{*chg}(T,x)=\sum x_i \Delta G^{*chg}_{i/S} \tag{4.11}$$

where *i/S* denotes solute i dissolved in solvent S and is given by

$$\Delta G^{*chg}_{i/S}=\Delta G^{*is}_{i}+\Delta G^{*cc}_{i}+\Delta G^{*res}_{i/S}+\Delta G^{*disp}_{i} \tag{4.12}$$

Each of the term was discussed in detail in Chapter 3. After obtaining a and b from the first principle calculation, the computation of isobaric VLLE follows the standard procedure (Prausnitz, Lichtenthaler, de Azevedo, 2004; Sandler, 1999). Justo-García, García-Sánchez, Díaz-Ramírez and Díaz-Herrera (2010) used SRK and PC-SAFT to compute the VLLE of a natural gas system. The problem was modelled by isothermal multi-phase flash problem. The general form of SRK (Soave, 1972) was used for this problem:

$$P=\frac{RT}{\nu-b}-\frac{a}{\nu(\nu+b)} \tag{4.13}$$

where a and b are calculated from critical properties (Justo-García et al., 2010). In PC-SAFT, molecules are divided into several spherical segments. The EoS is based on Helmholtz free energy of nonassociating chains. The attractive interactions are modelled separately. Details of PC-SAFT calculations are given in Section 2.7 of Chapter 2 (Alsaifi & Englezos, 2011).

Dividing molecules into segments is an attractive way to compute thermodynamic properties. Based on the polymer NRTL model (Chen, 1993), Chen and Crafts proposed the NRTL-segment activity coefficient (NRTL-SAC) model. Each component is conceptually assigned four segments, namely hydrophobic segment, polar attractive segment, polar-repulsive segment and hydrophilic segment. The combination of segments varies depending on the chemical structure of the molecules. The segment-based models are divided into two parts for calculations of the activity coefficient. They are termed as the residual part and the combinatorial part. The residual part originates from free energy change of solutes during solvation into a solvent. The combinatorial part takes care of interaction due to the shape and size of components. The logarithmic activity coefficient is the logarithmic sum of residual and combinatorial activity coefficients. The combinatorial part of the NRTL-SAC model is given by Equation 4.14.

$$\ln\gamma_i^C=\ln\left(\frac{\phi_i}{x_j}\right)+1-r_i\sum_j\frac{\phi_j}{x_j} \tag{4.14}$$

$$r_i = \sum_j r_{j,i} \tag{4.15}$$

$$\phi_i = \frac{r_i x_i}{\sum_j r_j x_j} \tag{4.16}$$

where:

x_i is the mole fraction of component i
$r_{j,i}$ is the number of segment j in component i
r_i is the total segment number in component i
φ_i is the segment mole fraction in the mixture.

The residual part of the activity coefficient is defined as

$$\ln \gamma_i^R = \ln \gamma_i^{lc} = \sum_m r_{m,i}\left(\ln \Gamma_m^{lc} - \ln \Gamma_m^{lc,i}\right) \tag{4.17}$$

where $\ln \Gamma_m^{lc}$ and $\ln \Gamma_m^{lc,i}$ are the activity coefficients of segment m in solution and in component i, respectively, defined by Equations 4.19 and 4.20:

$$\ln \Gamma_m^{lc} = \frac{\sum_j x_j G_{j,m} \tau_{j,m}}{\sum_k x_k G_{k,m}} + \sum_{m'} \frac{x_{m'} G_{mm'}}{\sum_k x_k G_{km'}} \left(\tau_{mm'} - \frac{\sum_j x_j G_{jm'} \tau_{jm'}}{\sum_k x_k G_{km'}} \right) \tag{4.18}$$

$$\ln \Gamma_m^{lc,l} = \frac{\sum_j x_{j,l} G_{jm} \tau_{jm}}{\sum_k x_{k,l} G_{km}} + \sum_{m'} \frac{x_{m',l} G_{mm'}}{\sum_k x_{k,l} G_{km'}} \left(\tau_{mm'} - \frac{\sum_j x_{j,l} G_{jm'} \tau_{jm'}}{\sum_k x_{k,l} G_{km'}} \right) \tag{4.19}$$

In Equations 4.18 and 4.19, l is referred to the component and j, k, m and m' are referred to the segments in each component. Segment-based mole fraction is denoted by $x_{j,l}$. The mole fractions of segments in the whole solution and in components are defined by Equations 4.20 and 4.21.

$$x_j = \frac{\sum_l x_l r_{j,l}}{\sum_z \sum_i x_z r_{j,z}} \tag{4.20}$$

$$x_{j,l} = \frac{r_{j,l}}{\sum_i r_{j,l}} \tag{4.21}$$

The local binary interaction parameter is given by Equation 4.22. As discussed in Chapter 2, $G_{i,j}$ is obtained from the genetic algorithm.

$$G_{i,j} = \exp(-\alpha_{i,j}\tau_{i,j}) \tag{4.22}$$

The first-principle-based COSMO-SAC model determines the liquid-phase nonideality using molecular interactions derived from the quantum chemical solvation calculation. Details of COSMO-SAC theory is given in Chapter 3. COSMO-SAC parameters are optimized against a large data set of experimental data and can be used for other types of systems without sacrificing accuracy. By this way, the model can be termed as a predictive model. One advantage of COSMO-based models over UNIFAC is that it can distinguish between isomers because molecules are drawn with respect to their conformal structures before performing optimization followed by COSMO calculation. Quantum chemical information in different conformers influences the activity coefficient calculations.

4.5 VLLE by Equilibrium and Flash Approach*

4.5.1 Equilibrium Equations

Consider a liquid mixture with overall composition z_i and molar holdup L which is at specified pressure P splits into two liquid phases (compositions are x_i^I and x_i^{II} molar holdups L^I and L^{II} respectively) that are at equilibrium with the vapor phase having composition y_i. At equilibrium all phases have the same chemical potential. So, overall and component material balances are

$$L = L^I + L^{II} \tag{4.23}$$

$$Lz_i = L^I x_i^I + L^{II} x_i^{II} \tag{4.24}$$

Defining the distribution coefficient as K_i and LLE split of phase I as ψ, $\psi = L^I/L$. As the two liquid phases are in phase equilibrium with each other and to the vapor phase, thus we have

$$y_i P = x_i^I \gamma_i^I P_i^{\text{sat}} = x_i^{II} \gamma_i^{II} P_i^{\text{sat}} \tag{4.25}$$

$$x_i^I \gamma_i^I P_i^{\text{sat}} = x_i^{II} \gamma_i^{II} P_i^{\text{sat}}$$

$$\frac{x_i^I}{x_i^{II}} = \frac{\gamma_i^{II}}{\gamma_i^I} = K_i \tag{4.26}$$

* Sections 4.5 and 4.6 reprinted (adapted) with permission from D. Kundu, T. Banerjee, Multicomponent vapor–liquid–liquid equilibrium prediction using an a priori segment-based model, *Ind. Eng. Chem. Res.* 50(2011) 14090–14096, 2011. Copyright 2011 American Chemical Society.

Here γ_i^{II} and γ_i^{I} (where i = number of compounds in a single phase, that is, organic/aqueous/vapor) are calculated from the COSMO-SAC model. Thus, the mole fractions in the vapor phase and the liquid phase are given by

$$x_i^I = \frac{K_i z_i}{\left[1+\psi\left(K_i-1\right)\right]};\ x_i^{II} = \frac{z_i}{\left[1+\psi\left(K_i-1\right)\right]} \quad \text{and}$$
$$y_i = \frac{x_i^I \gamma_i^I P_i^{\text{sat}}}{P} \quad \text{or} \quad y_i = \frac{x_i^{II} \gamma_i^{II} P_i^{\text{sat}}}{P} \tag{4.27}$$

ψ is calculated by modified Rachford–Rice (Seader & Henley, 2007) equation (Equation 4.28)

$$\sum_{i=1}^{n} \frac{z_i\left(1-K_i\right)}{\left[1+\psi\left(K_i-1\right)\right]} = 0 \tag{4.28}$$

Additionally, the following three constraints are also to be satisfied

$$\sum_{i=1}^{n} x_i^I = 1,\ \sum_{i=1}^{n} x_i^{II} = 1,\ \sum_{i=1}^{n} y_i = 1 \tag{4.29}$$

Prediction by the equilibrium approach is done by Rachford–Rice algorithm. The algorithm is described by the following flowchart (Figure 4.1).

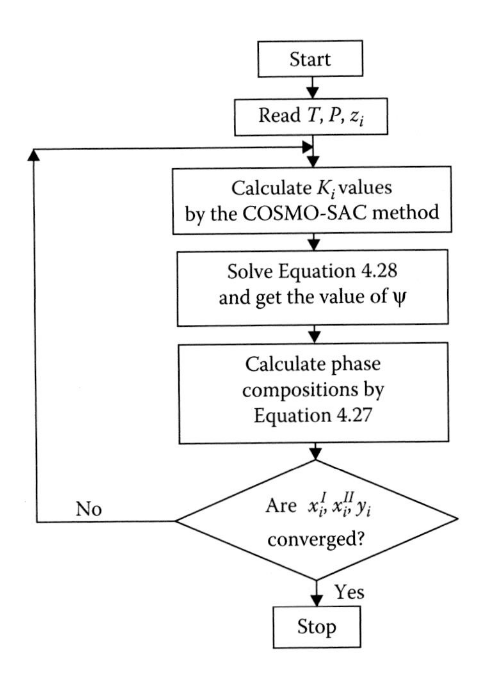

FIGURE 4.1
Flowchart of modified Rachford–Rice algorithm.

4.5.2 Flash Equations

In a dynamic simulation of VLLE, the equation-solving approach is typically referred to a positive flash. A set of nonlinear algebraic equations are deduced from mass balances and equilibrium relations and later stability analysis is performed. Physical insights into multi-phase calculation are incorporated in the calculation. The concept of negative flash is introduced by Neochill and Chambrette, and Withson and Michelsen (Neoschil & Chambrette, 1978; Whitson & Michelsen, 1989) for VLE. The physical domain of phase split between 0 and 1 was extended to outside the domain of phase split. The physical meaning of phase split having outside 0 and 1 is the existence of a single phase (Figure 4.2).

Henley and Rosen (Seader & Henley, 2007) presented a rigorous three-phase flash calculation scheme which is an extension of conventional two-phase flash. The simplified version of Henley–Rosen scheme was presented by Sampath and Leipziger (1985) and is applicable to the hydrocarbon phase (denoted by I) and the aqueous phase (denoted by II). The vapor phase is denoted by V. The method is described here briefly. The three constraints (Equation 4.29) mentioned above are to be satisfied here too. α is the ratio of the amount of vapor to the total feed and β is the liquid phase split which are defined as

$$\alpha = \frac{V}{F} \quad \text{and} \quad \beta = \frac{L^I}{L^I + L^{II}}$$

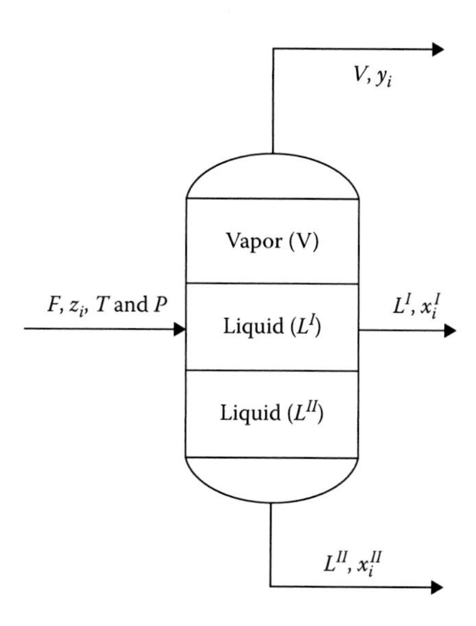

FIGURE 4.2
Schematic diagram for three-phase flash system.

Overall and component material balances are given by

$$F = V + L^I + L^{II} \tag{4.30}$$

$$z_i = \beta(1-\alpha)x_i^I + (1-\alpha)(1-\beta)x_i^{II} + \alpha y_i \tag{4.31}$$

Individual phase compositions can be computed from the following equations:

$$x_i^I = \frac{z_i}{\left[\beta(1-\alpha)+(1-\alpha)(1-\beta)\dfrac{K_i^I}{K_i^{II}} + \alpha K_i^I\right]} \tag{4.32}$$

$$y_i = K_i^I x_i^I \tag{4.33}$$

$$x_i^{II} = \frac{K_i^I}{K_i^{II}} x_i^I \tag{4.34}$$

where

$$K_i^I = \frac{\gamma_i^I P_i^{\text{sat}}}{P} \quad \text{and} \quad K_i^{II} = \frac{\gamma_i^{II} P_i^{\text{sat}}}{P}$$

Here, γ_i^{II} and γ_i^I (where i = number of compounds in a single phase, that is, organic/aqueous/vapor) are again calculated from the COSMO-SAC model. The criteria for VLLE can be written as

$$\sum_{i=1}^{n} x_i^I - \sum_{i=1}^{n} y_i = 0 \tag{4.35}$$

$$\sum_{i=1}^{n} x_i^{II} - \sum_{i=1}^{n} y_i = 0 \tag{4.36}$$

$$\sum_{i=1}^{n} x_i^I - \sum_{i=1}^{n} x_i^{II} = 0 \tag{4.37}$$

For three phases to coexist $0 < \alpha < 1$ and $0 < \beta < 1$ must be satisfied.

Sampath and Leipziger (1985) did flash prediction using a simplified algorithm of Henley–Rosen method. As none of the systems are binary in the present work, the binary loop of the original algorithm was modified. Newton–Raphson convergence scheme was applied making an assumption that the activity of water in the aqueous phase is 1, that is, $\alpha_{H_2O}^{II} = (\gamma x)_{H_2O}^{II} = 1$.

In many hydrocarbon–water systems, the mole fraction of water in the aqueous liquid phase is close to unity. If the pure liquid is chosen as a reference state for water in the aqueous liquid phase, the activity coefficient of water is close to unity. With this assumption, using Equations 4.32 and 4.34, β becomes a function of α, that is,

$$\beta = \frac{z_{H_2O}\gamma_{H_2O}^{II} - \alpha\left(K_{H_2O}^{II} - 1\right) - 1}{(1-\alpha)\left(\dfrac{K_{H_2O}^{II}}{K_{H_2O}^{I}} - 1\right)} \tag{4.38}$$

Good initial estimates are necessary to avoid a trivial solution. A good estimate is to take $\alpha = 0.33$ (Sampath & Leipziger, 1985), whereby β is then calculated from Equation 4.38. The flash algorithm is described following a flowchart (Figure 4.3) and is applied in Systems 6 and 7 of Table 4.1. The goodness of fit is usually gauged by the root-mean-square deviation (RMSD), which is defined by (Equation 4.39)

$$\text{RMSD}(\%) = 100\left[\sum_{k=1}^{m}\sum_{i=1}^{c}\sum_{j=1}^{p}\frac{\left(x_{\text{cal},ik}^{j} - x_{\text{expt},ik}^{j}\right)^2}{pmc}\right]^{1/2} \tag{4.39}$$

where:

m refers to the number of tie lines
c is the number of components
p is the number of phases

4.6 Problem Formulation and Predicted Phase Equilibria of VLLE

Isobaric VLE and VLLE data of the water–ethanol–hydrocarbon system is important to design the distillation column for the dehydrogenation of ethanol. Ethanol–gasoline blends are particularly important because alcohol increases the octane level and promotes complete fuel burning (Gomis, Font, & Saquete, 2006). Small amount of water causes harmful effect in end application; therefore, different entrainers are used to dehydrate it prior to blending with gasoline. Traditionally, benzene is used as an entrianer in the heterogeneous azeotropic distillation of a water–ethanol system. Due to the carcinogenic effect of benzene, other entrainers are also investigated. Here, we consider experimental data involving hexane, heptane, cyclohexane, toluene and isooctane as entrainers to water–ethanol systems. Diisopropyl ethers

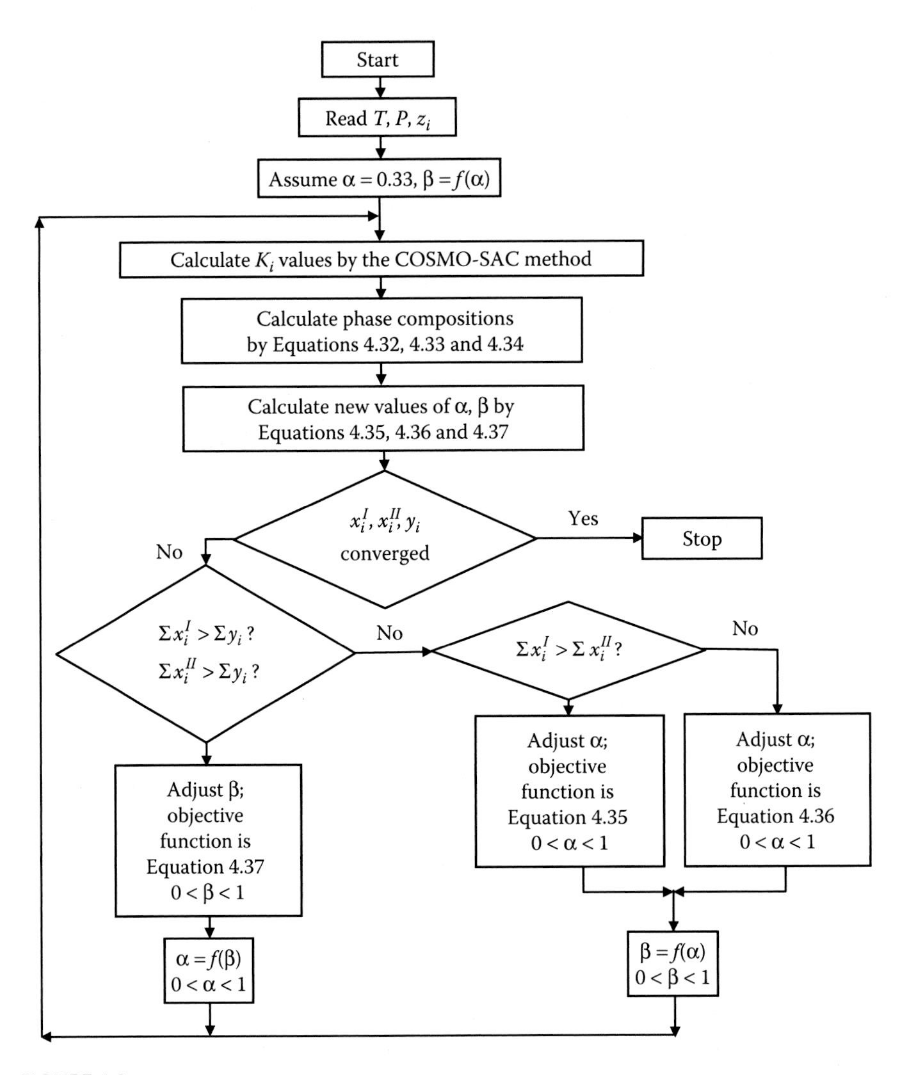

FIGURE 4.3
Flow chart of Henley–Rosen algorithm with Sampath and Leipziger approximation.

and di-n-propyl ethers are other important additives used to enhance octane number for gasoline. Thus, VLLE data of two ether–alcohol–water containing systems (Lladosa, Montón, Burguet, & de la Torre, 2008) were taken into consideration here. We took an equilibrium data of ternary water–alcohol systems (Asensi, Moltó, del Mar Olaya, Ruiz, & Gomis, 2002) as a flagship of hydrogen-bond-containing systems. The aim to consider this system is whether predictive models like COSMO-SAC can efficiently handle hydrogen-bond-containing systems. Gomis and coworkers (Asensi et al., 2002; Gomis, Font, Pedraza, & Saquete, 2005, 2007; Gomis, Font, & Saquete, 2008; Gomis et al., 2000, 2006; Lladosa et al., 2008; Pequenín, Asensi, & Gomis, 2010) did a series

TABLE 4.1

List of VLLE Systems Studied

Sr. No.	System	Type	Reference
1	Water + ethanol + hexane	Ternary	(Gomis et al., 2007)
2	Water + ethanol + heptane	Ternary	(Gomis et al., 2006)
3	Water + ethanol + cyclohexane	Ternary	(Gomis et al., 2005)
4	Water + ethanol + toluene	Ternary	(Gomis et al., 2008)
5	Water + 1-propanol + 1-pentanol	Ternary	(Asensi et al., 2002)
6	Water + 1-propanol + di-n-propyl ether	Ternary	(Lladosa et al., 2008)
7	Water + 2-propanol + diisopropyl ether	Ternary	(Lladosa et al., 2008)
8	Water + ethanol + cyclohexane + isooctane	Quaternary	(Pequenín et al., 2010)

of VLLE experiments of ternary and quaternary systems. The experimental data are usually compared with NRTL, UNIQUAC and UNIFAC models. All the experiments are performed at a pressure of 1 atmosphere and a short temperature difference (~5 K). Lladosa et al. (2008) did the VLLE experiment for water–alcohol–ether systems at 1 atmosphere pressure. These systems are reported in Table 4.1 for which the predictions of phase equilibria were carried out by the equilibrium approach and the flash approach.

The initial guess of feed mole fractions is taken as the average mole fraction of the three phases of the component in consideration. Saturation pressure is calculated by Antoine equation. Antoine constants are reported in Table 4.2.

Antoine equation is given by

$$\log_{10} P^{sat} = A - \frac{B}{t + C} \tag{4.40}$$

where:

P^{sat} is the saturation pressure in mmHg

t is °C

The prediction of phase equilibria involves the calculation of activity coefficients for every compound in both liquid phases. COSMO-SAC calculation was done to predict activity coefficients. The parameters for COSMO-SAC are given in Table 4.3. The representative outcome of COSMO calculation is σ-profile that is a two-dimensional representation of three-dimensional charge distribution among molecules. The representative σ-profile of water + ethanol + hexane is shown in Figure 4.4. From σ-profile, it is evident that hexane is nonpolar, that is, its peak lies between −0.0084 e/Å^2 and 0.0084 e/Å^2. But ethanol and water have segments lying in nonpolar as well as hydrogen bond donor and acceptor regions.

Predicted tie lines of Systems 1, 2 and 3 of Table 4.1 and Figure 4.5a–c have fewer slopes as compared to experimental tie lines. At the lower part of the phase envelope, water molecules mainly contribute to hydrogen bonding as compared to ethanol. The presence of more amount of ethanol as compared

TABLE 4.2
Antoine Constants

Name of Compounds	A	B	C
Hexane	6.91058	1189.64	226.28
Heptane	6.89386	1264.37	216.64
Cyclohexane	6.85146	1206.47	223.136
Toluene	6.95087	1342.31	219.187
Isooctane	6.80304	1252.59	220.119
Ethanol	8.11220	1592.864	226.184
1-propanol	8.37895	1788.02	227.438
2-propanol	8.87829	2010.33	252.636
1-pentanol	7.39824	1435.57	179.798
Di-n-propyl ether	6.94760	1256.50	219.00
Diisopropyl ether	6.84953	1139.34	218.742
Water	8.07131	1730.63	233.426

Source: Rao, Y. V. C. *Chemical Engineering Thermodynamics*, Hyderabad, University Press (India) Pvt. Ltd, 2003.

TABLE 4.3
COSMO-SAC Parameters

Name	Value	Unit
Effective area (a_{eff})	7.5	Å^2
Cut-off value for hydrogen bonding interaction (c_{hb})	0.0084	e/Å^2
Constant for hydrogen bonding interaction (σ_{hb})	85580	(Kcal/mol)(Å^4/e^2)

Source: Lin, S.-T. and Sandler, S. I. *Ind. Eng. Chem. Res.*, 41, 899–913, 2002.

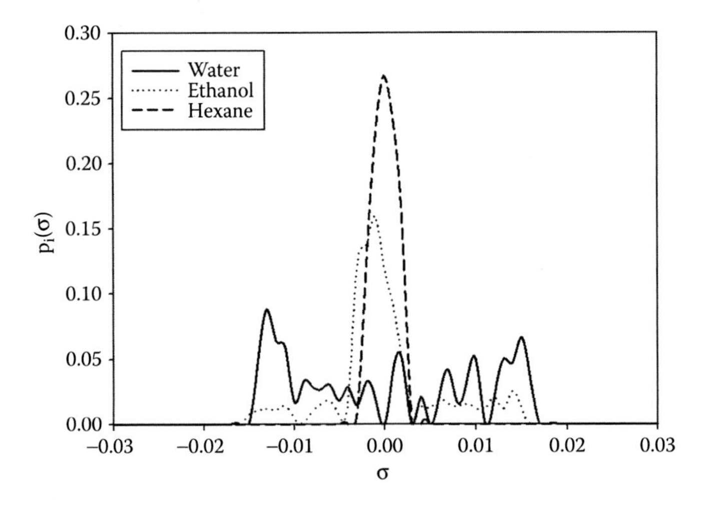

FIGURE 4.4
σ-profile of water + ethanol + hexane. (σ in e/Å^2).

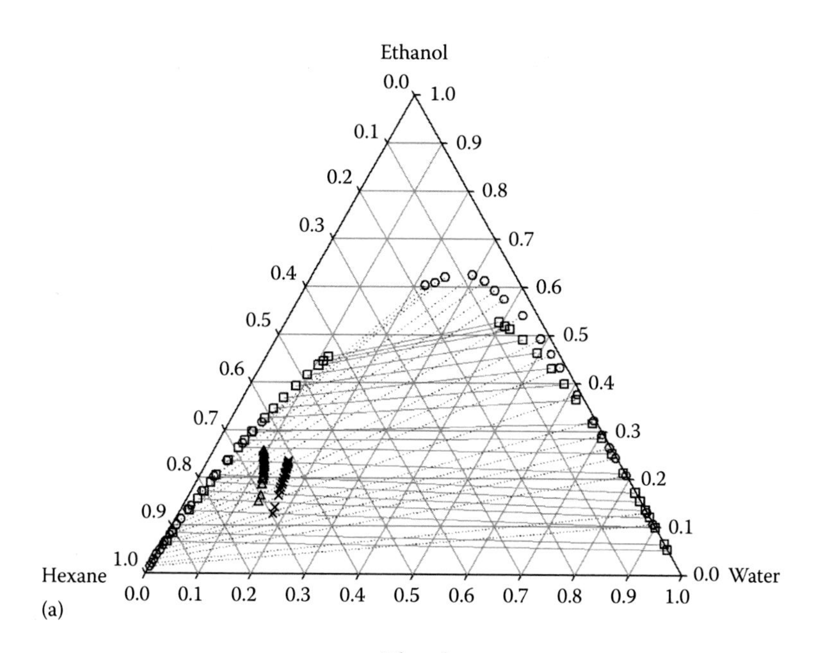

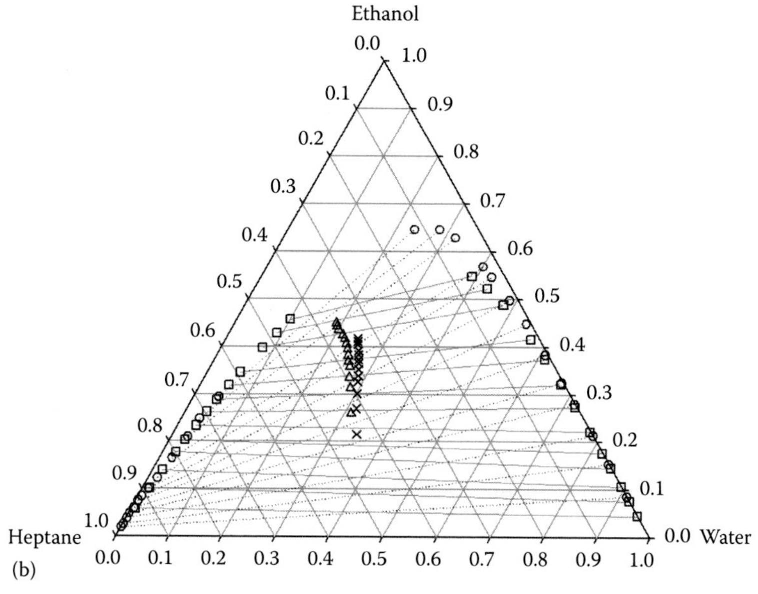

FIGURE 4.5
Comparison of VLLE from experiments and predictions for (a) water + ethanol + hexane, (b) water + ethanol + heptane, *(Continued)*

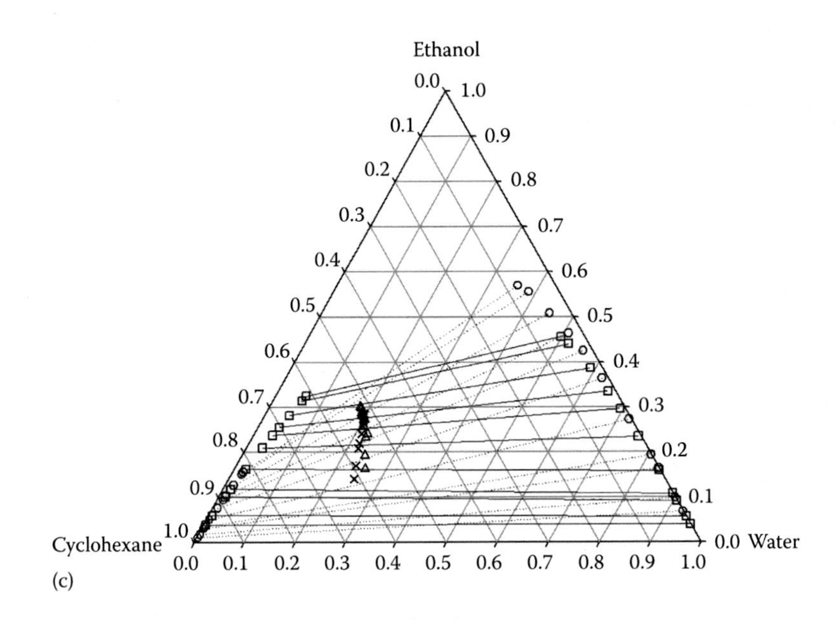

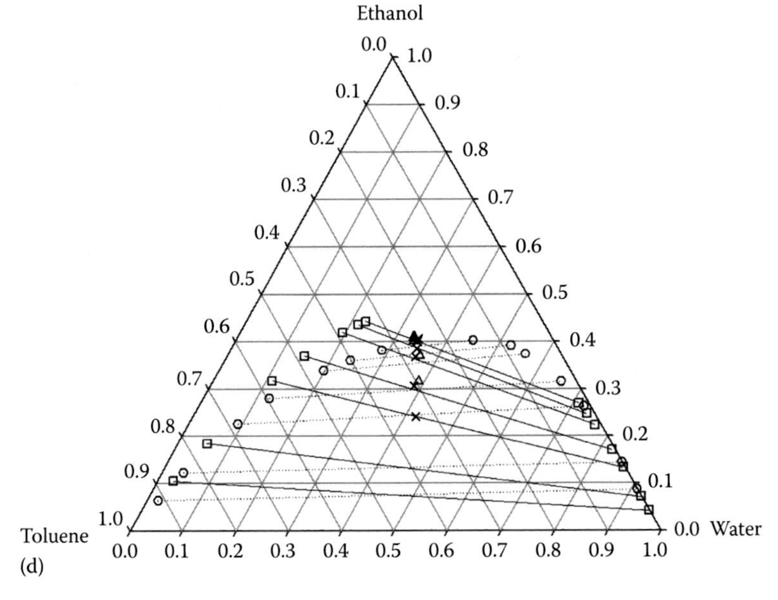

FIGURE 4.5 (Continued)
Comparison of VLLE from experiments and predictions for (c) water + ethanol + cyclohexane, (d) water + ethanol + toluene, *(Continued)*

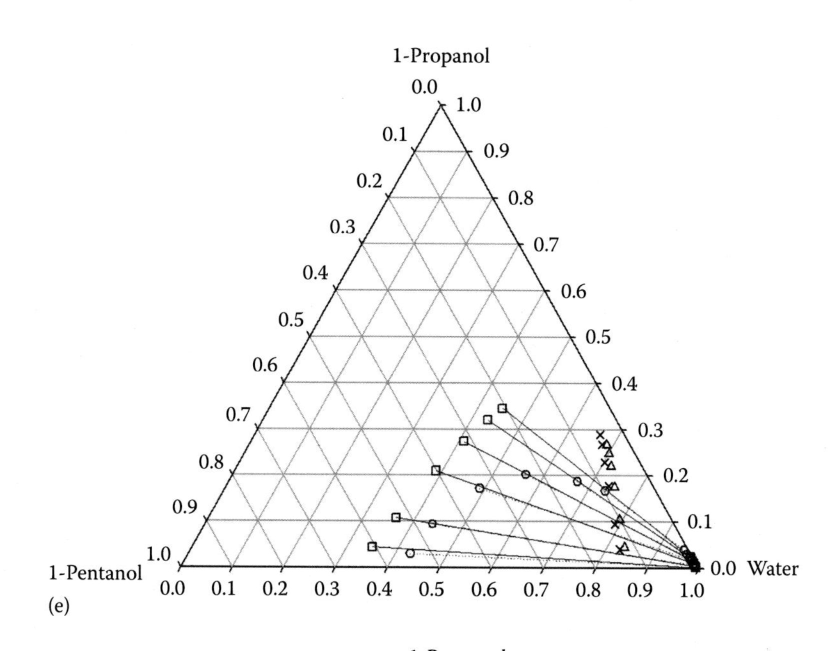

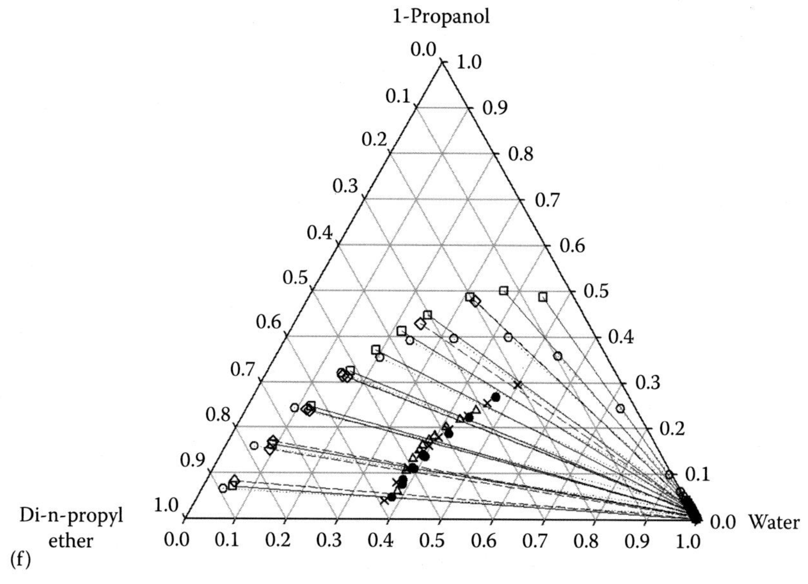

FIGURE 4.5 (Continued)
Comparison of VLLE from experiments and predictions for (e) water + 1-propanol + 1-pentanol, (f) water + 1-propanol + di-n-propyl ether, *(Continued)*

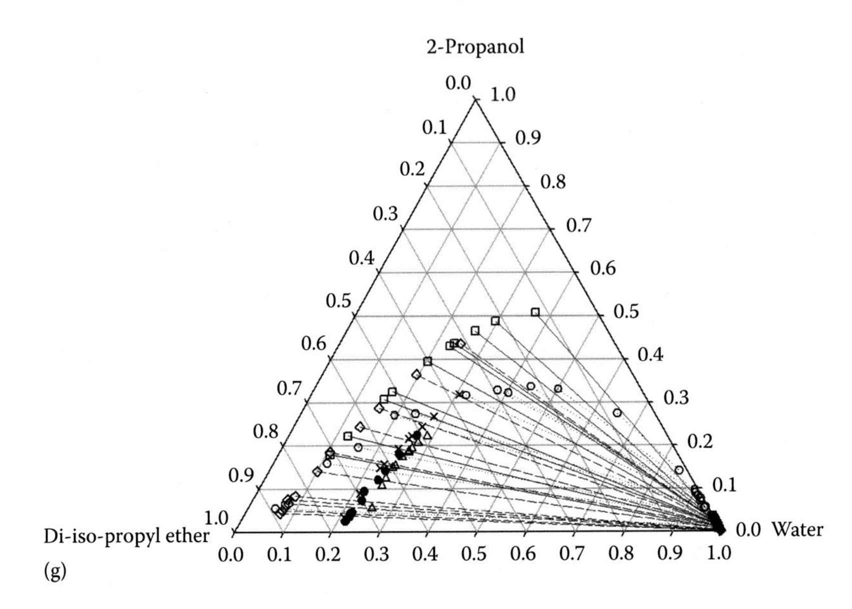

FIGURE 4.5 (Continued)
Comparison of VLLE from experiments and predictions for (g) water + 2-propanol + diisopropyl ether. N.B. For all the ternary diagrams the following notations will be used: (○) liquid phase experimental data, (Δ) vapor phase experimental data, (□) liquid phase predicted by equilibrium relation, (×) vapor phase predicted by equilibrium relation, (◇) liquid phase predicted by flash relation and (●) vapor phase predicted by flash relation.

to water in the organic phase results in higher predicted points in the organic phase than the corresponding experimental values. The deviations are low near the base of the aqueous phase where the concentrations of ethanol and hexane/heptane/cyclohexane are very low. As the concentrations of ethanol and water become equal or come close to each other, the deviation increases. This is seen at the upper part of the ternary diagram in Figure 4.5a–c. The trend in the organic phase is maintained even when ethanol concentration is increasing and the corresponding solvent concentration is decreasing giving an RMSD of 14.59%, 14.23% and 11.17%, respectively. The deviation in the aqueous phase towards the plait point is higher. Near the plait point, the composition of water is almost half to that of ethanol; thus, ethanol is responsible for contributing towards hydrogen bonding. RMSD in the aqueous phase for the three systems are 12.38%, 11.77% and 8.93%, respectively. In the original COSMO-SAC model, hydrogen bonding is not correctly accounted for when the predicted mole fractions of water and ethanol are closer. The hydrogen bonding correction for COSMO-SAC will be discussed in Chapter 5. These lower the area of the binodal curve as compared to the experimental phase envelope.

For System 4 of Table 4.1 and Figure 4.5d, in the aqueous phase, the mole fraction of water is always greater than ethanol. Hydrogen bonding is mainly contributed by water; thus, predicted mole fraction of water is much higher, while that of ethanol is lower. The organic phase follows the same trend as observed in previous three systems. Thus, the concentrations of ethanol and water make the tie line slope negative, although it is positive by experimental data. Nevertheless, COSMO-SAC predicts a closer phase envelope corresponding to experimental tie lines. NRTL, UNIQUAC and UNIFAC predict much bigger phase envelopes (Gomis et al., 2008).

In System 5 of Table 4.1 and Figure 4.5e, in the aqueous phase, the composition of water is closer to unity, thus making contribution in hydrogen bonding resulting in accurate prediction with an RMSD of only 0.76%. Predicted tie lines merge with the corresponding experimental tie lines but due to the presence of three similar polar components, organic phase compositions vary (RMSD of 12.57%), resulting in a bigger phase envelope. Hydrogen bonding contribution will be visible here most as all the compounds (water, 1-propanol and 1-pentanol) have hydrogen bonding acceptor segments (due to the presence of oxygen atoms) and hydrogen bonding donor segments (due to the presence of hydrogen atoms bonded with oxygen atom). Both cross and self-hydrogen bonding are visible in this system. The mole fraction of water in the aqueous phase is near to unity for all the tie lines making water a dominant contributor of hydrogen bonding in this phase. As a result mainly cross hydrogen bonding is visible here and RMSD is very less. In the organic phase, three species are present with comparable mole fractions. Thus, hydrogen bonding is present among all type of species. The present method of hydrogen bonding calculation cannot take into account all such combinations and give higher RMSD.

Systems 6 and 7 of Table 4.1 and Figure 4.5f and g have ethers (di-n-propyl ether and diisopropyl ether, respectively) as solvents. The ternary diagrams of these systems are shown below.

Ether's σ-profile lies in the nonpolar region as well as in the hydrogen bond acceptor region (Figure 4.6).

The aqueous phase prediction via the equilibrium approach is somewhat accurate. The RMSD calculated using the equilibrium approach for Systems 6 and 7 for the aqueous phase are 2.04% and 4.72%, respectively, whereas using the flash approach are 2.45% and 5.61%, respectively. In the organic phase, at lower mole fraction of ether (or higher mole fraction of water), the predicted data (both by equilibrium approach and flash approach) are near the experimental points, but at higher ether mole fraction, deviations are bigger. Because total polar segments of ether are in the acceptor region, then at comparable mole fractions, that is, near the plait point, there will be a deficiency of donor segments. But COSMO-SAC cannot take into account these extra segments and generate the same amount of donor and acceptor segments. For that reason, the phase envelope predicted by the equilibrium approach is slightly higher than that predicted by the flash approach, although the RMSD in the organic phase by the equilibrium approach (8.7% and 10.66%) are much lower than the flash approach (14.06% and 24.27%). The split values, α and β, of these systems are reported in Table 4.4. It is seen that both parameters lie between 0 and 1 which implies Equations 4.35 through 4.37 are satisfied.

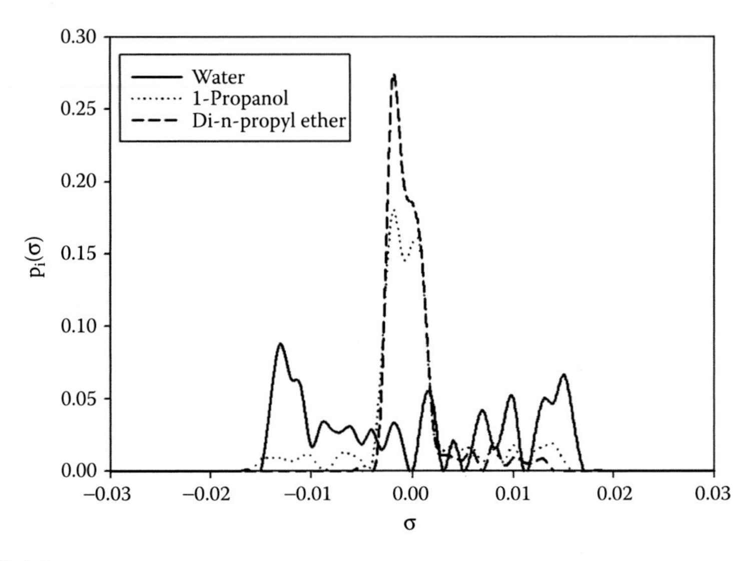

FIGURE 4.6
σ-profile of water + 1-propanol + di-n-propyl ether (σ in e/Å²).

TABLE 4.4
Splits for Systems 6 and 7

System	α	β
Water + 1-propanol + di-n-propyl ether (System 6)	0.3498	0.8927
	0.3364	0.5008
	0.3322	0.5024
	0.3148	0.5066
	0.2981	0.5102
	0.2445	0.5201
	0.2973	0.5154
	0.2685	0.5195
	0.2668	0.5177
	0.3031	0.5091
Water + 2-propanol + diisopropyl ether (System 7)	0.3376	0.5025
	0.3297	0.5066
	0.3208	0.5105
	0.3130	0.5140
	0.2826	0.5269
	0.2234	0.5498
	0.1924	0.5604
	0.1418	0.5764
	0.1237	0.5816
	0.3055	0.5200
	0.3218	0.5105

The quaternary system was plotted via the pseudo-ternary (cyclohexane and isooctane are taken together) approach in four different ternary diagrams (Figure 4.7a–d) as described in literature (Pequenín et al., 2010). As both cyclohexane and isooctane are nonpolar, hydrogen bonding characteristics in both

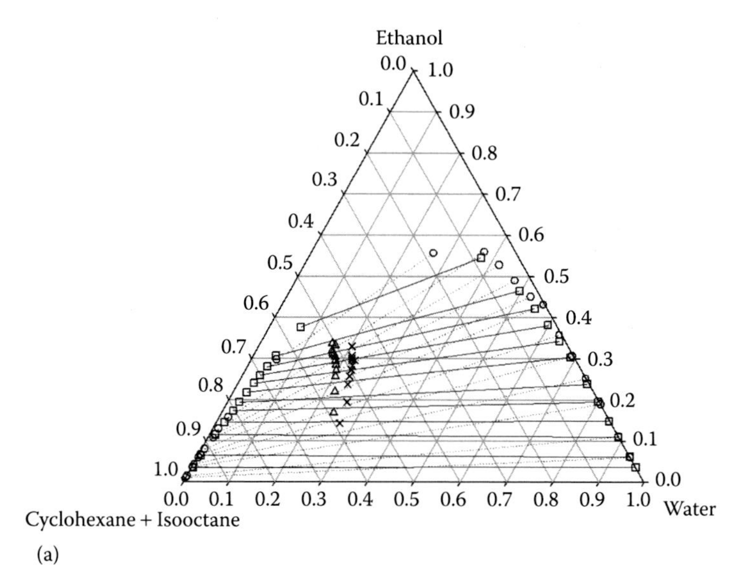

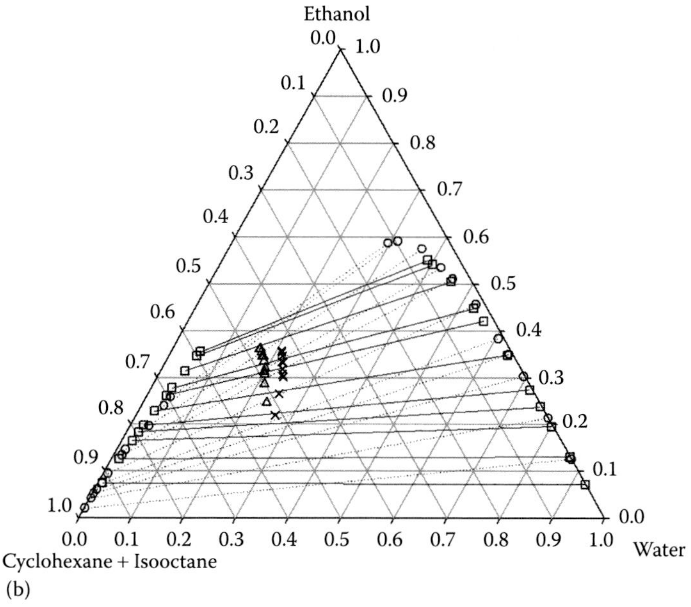

FIGURE 4.7
Comparison of pseudo-ternary VLLE from experiments and predictions for water + ethanol + cyclohexane + isooctane. (a) $M = 0.2$, (b) $M = 0.4$. *(Continued)*

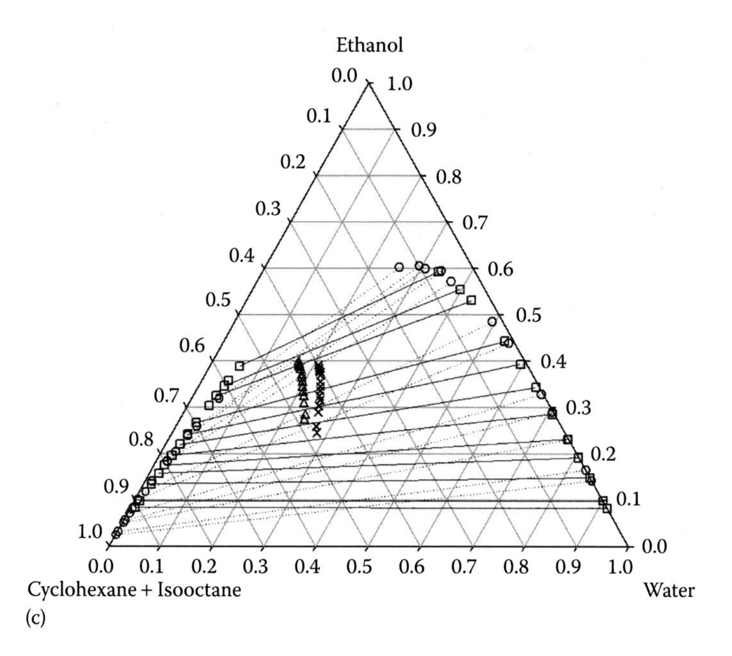

(c)

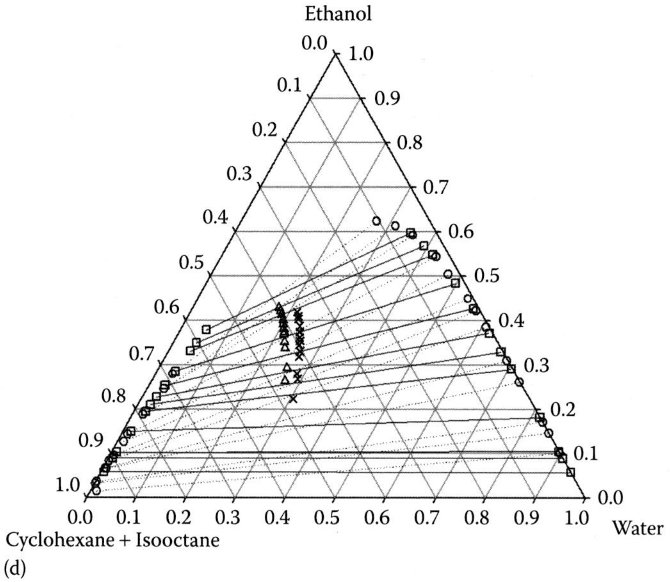

(d)

FIGURE 4.7 (Continued)
Comparison of pseudo-ternary VLLE from experiments and predictions for water + ethanol + cyclohexane + isooctane. (c) $M = 0.6$ and (d) $M = 0.8$.

liquid phases follow Systems 1, 2 and 3 of Table 4.1. The prediction of a phase envelope agrees well with experimental points, with the RMSD of 7.09% in the organic phase and 6.23% in the aqueous phase. This is smaller than that of Systems 1, 2 and 3 of Table 4.1. In all the cases, vapor phase deviations are less than 7%.

Quaternary systems are plotted using the following notation: M = 0.2, 0.4, 0.6, 0.8 are reported in literature (Pequenín et al., 2010). (○) liquid phase experimental data, (Δ) vapor phase experimental data, (□) liquid phase predicted by equilibrium relation and (×) vapor phase predicted by equilibrium relation.

$$M = \frac{\text{Mole fraction of isooctane}}{\text{Mole fraction of cyclohexane} + \text{Mole fraction of isooctane}}$$

The RMSD values are tabulated in Tables 4.5 and 4.6. While Table 4.5 shows the RMSD values for all the eight systems predicted via the equilibrium relation, Table 4.6 gives the RMSD values for Systems 6 and 7 via the flash relation. In both predictions, average RMSD in the organic phase (10.67% for equilibrium- and 19.17% for flash-based algorithm) is much higher as compared to the aqueous phases, 7.45% (equilibrium based) and 4.03% (flash based) and the vapor phases, 3.26% (equilibrium based) and 5.84% (flash based), indicating that there are interactions in the organic phase.

The ternary water–ethanol–entrainer systems form an azeotrope inside the distillation column. From the experimental isothermal bimodal curve and the vapor line, a ternary azeotrope is clearly visible for hexane, heptane and cyclohexane entrainers. For water–ethanol–hexane system (System 1), the azeotrope appears between 13th and 14th tie lines because from 1st to 13th tie lines, the vapor lines come above the liquid–liquid tie line and for 14–20th, it comes under the liquid–liquid tie line. From experimental data, the vapor phase coincides with the liquid tie line between 13th and 14th tie lines and an azeotrope is formed. From the equilibrium approach of VLLE computation, the azeotrope formation appears at the 6th tie line. VLLE computation predicts a bigger phase envelope but undermines the azeotropic composition. The predicted composition for a ternary azeotrope is x_1 (water) = 0.1642, x_2 (ethanol) = 0.1870 and x_3 (hexane) = 0.6489. Liquid composition for the water-rich phase is x_1 (water) = 0.8439, x_2 (ethanol) = 0.1548 and x_3 (hexane) = 0.0013 and for the organic phase is x_1 (water) = 0.0286, x_2 (ethanol) = 0.1909 and x_3 (hexane) = 0.7805. For water–ethanol–heptane system (System 2), the azeotrope formation exactly mimics the experimental data and lies within 10th and 11th tie lines. From numerical interpolation, the composition is determined as

TABLE 4.5

RMSD by Equilibrium Prediction* (Equations 4.28 and 4.29)

Phase	System 1	System 2	System 3	System 4	System 5	System 6	System 7	System 8	Average RMSD
Organic	14.59	14.23	11.17	6.34	12.57	8.70	10.66	7.09	10.67
Aqueous	12.38	11.77	8.93	12.75	0.76	2.04	4.72	6.23	7.45
Vapor	3.39	6.58	1.54	2.90	1.37	3.50	4.15	2.63	3.26
Overall	11.22	11.32	8.30	8.39	7.31	5.54	7.14	5.66	8.11

*Systems 1–8 are as reported in Table 4.1.

TABLE 4.6

RMSD (%) by Flash Prediction (Equations 4.31 through 4.38)

Phase	System 6*	System 7*	Average RMSD
Organic	14.06	24.27	19.17
Aqueous	2.45	5.61	4.03
Vapor	3.31	8.36	5.84
Overall	8.46	15.17	11.82

*Systems 6 and 7 are as reported in Table 4.1.

x_1 (water) = 0.2545, x_2 (ethanol) = 0.3975 and x_3 (heptane) = 0.3479, with the aqueous phase composition as x_1 (water) = 0.5166, x_2 (ethanol) = 0.4544 and x_3 (heptane) = 0.0290 and the organic phase composition x_1 = 0.0711, x_2 = 0.3684 and x_3 = 0.5605. From the experimental VLLE data of water–ethanol–cyclohexane system (system 3), the azeotrope lies between 9th and 10th tie lines (Gomis et al., 2005). However, the computational approach undermines the experimental findings and the azeotrope lies at the 8th tie line with a composition of x_1 (water) = 0.1945, x_2 (ethanol) = 0.2797 and x_3 (cyclohexane) = 0.5258. A homogeneous azeotrope is formed for water–ethanol–toluene system (System 4) at the 7th tie line which is very near to the plait point. The composition at the azeotrope is determined at x_1 (water) = 0.3438, x_2 (ethanol) = 0.4066 and x_3 (toluene) = 0.2496. No azeotropic distillation is observed for water–1-propanol–1-pentanol and water–alcohol–ether systems (System 5, 6 and 7). Quaternary systems are reported as pseudo-ternary systems combining cyclohexane and octane. Thus, the pseudo-ternary diagram will not give hints of the formation of azeotropes. The tetrahedral representation of quaternary VLLE data reveals that equilibrium vapor phases lie outside sectional planes containing liquid phases (Pequenín et al., 2010). Thus, this system does not form a heterogeneous azeotrope. A complete list of the predicted composition of the azeotropic mixture and the liquid composition at the organic and the aqueous phases is given in Table 4.7.

Because VLLE is important for designing chemical equipment and because experiments are time consuming, a predictive method is necessary to describe the phase behavior of azeotrope-forming systems. The COSMO-SAC model predicts activity coefficients and phase compositions based on a quantum chemical coupled with statistical mechanical calculations. The resulting phase envelope follows a similar trend as observed via the NRTL, the UNIQUAC and the UNIFAC models. The average RMSD obtained for all the eight systems using the equilibrium approach was 8.11%. In the same manner, the overall RMSD obtained for two systems using flash calculation is 11.81%.

TABLE 4.7

Comparison between Predicted and Experimental Azeotropic Compositions

		Predicted Data				Experimental Data			
	Components	Azeotropic Composition	Organic Phase Composition	Aqueous Phase Composition	Temperature (K)	Azeotropic Composition	Organic Phase Composition	Aqueous Phase Composition	Temperature (K)
System 1	Water (x_1)	0.1642	0.0286	0.8439	329.30	0.1050	0.0160	0.4010	329.21
	Ethanol (x_2)	0.1870	0.1909	0.1548		0.2360	0.1380	0.0160	
	Hexane (x_3)	0.6489	0.7805	0.0013		0.6580	0.8470	0.1380	
System 2	Water (x_1)	0.2545	0.0711	0.5166	341.91	0.2050	0.0300	0.3410	341.83
	Ethanol (x_2)	0.3975	0.3684	0.4544		0.4320	0.1950	0.6140	
	Heptane (x_3)	0.3479	0.5605	0.0290		0.3630	0.7750	0.0450	
System 3	Water (x_1)	0.1945	0.0422	0.6507	335.63	0.1880	0.0180	0.4330	335.54
	Ethanol (x_2)	0.2797	0.2558	0.3343		0.2920	0.1330	0.5210	
	Cyclohexane (x_3)	0.5258	0.7020	0.0150		0.5200	0.8490	0.0460	
System 4	Water (x_1)	0.3438	0.2268	0.7115	347.64	0.3320	NA		347.6
	Ethanol (x_2)	0.4066	0.4419	0.2696		0.412			
	Toluene (x_3)	0.2496	0.3313	0.0189		0.256			

References

Abrams, D. S., & Prausnitz, J. M. (1975). Statistical thermodynamics of liquid mixtures: A new expression for the excess Gibbs energy of partly or completely miscible systems. *AIChE Journal, 21*(1), 116–128. doi:10.1002/aic.690210115.

Alsaifi, N. M., & Englezos, P. (2011). Prediction of multiphase equilibrium using the PC-SAFT equation of state and simultaneous testing of phase stability. *Fluid Phase Equilibria, 302*(1–2), 169–178. doi:10.1016/j.fluid.2010.09.002.

Asensi, J. C., Moltó, J., del Mar Olaya, M. A., Ruiz, F., & Gomis, V. (2002). Isobaric vapour–liquid equilibria data for the binary system 1-propanol+1-pentanol and isobaric vapour–liquid–liquid equilibria data for the ternary system water+1-propanol+1-pentanol at 101.3 kPa. *Fluid Phase Equilibria, 200*(2), 287–293. doi:10.1016/S0378-3812(02)00040-7.

Chen, C.-C. (1993). A segment-based local composition model for the Gibbs energy of polymer solutions. *Fluid Phase Equilibria, 83,* 301–312. doi:10.1016/0378-3812(93)87033-W.

Fredenslund, A., Jones, R. L., & Prausnitz, J. M. (1975). Group-contribution estimation of activity coefficients in nonideal liquid mixtures. *AIChE Journal, 21*(6), 1086–1099. doi:10.1002/aic.690210607.

Gomis, V., Font, A., Pedraza, R., & Saquete, M. D. (2005). Isobaric vapor–liquid and vapor–liquid–liquid equilibrium data for the system water + ethanol + cyclohexane. *Fluid Phase Equilibria, 235*(1), 7–10. doi:10.1016/j.fluid.2005.07.015.

Gomis, V., Font, A., Pedraza, R., & Saquete, M. D. (2007). Isobaric vapor–liquid and vapor–liquid–liquid equilibrium data for the water–ethanol–hexane system. *Fluid Phase Equilibria, 259*(1), 66–70. doi:10.1016/j.fluid.2007.04.011.

Gomis, V., Font, A., & Saquete, M. D. (2006). Vapour–liquid–liquid and vapour–liquid equilibrium of the system water + ethanol + heptane at 101.3 kPa. *Fluid Phase Equilibria, 248*(2), 206–210. doi:10.1016/j.fluid.2006.08.012.

Gomis, V., Font, A., & Saquete, M. D. (2008). Homogeneity of the water + ethanol + toluene azeotrope at 101.3 kPa. *Fluid Phase Equilibria, 266*(1–2), 8–13. doi:10.1016/j.fluid.2008.01.018.

Gomis, V., Ruiz, F., & Asensi, J. C. (2000). The application of ultrasound in the determination of isobaric vapour–liquid–liquid equilibrium data. *Fluid Phase Equilibria, 172*(2), 245–259. doi:10.1016/S0378-3812(00)00380-0.

Hsieh, C.-M., & Lin, S.-T. (2009a). First-principles predictions of vapor–liquid equilibria for pure and mixture fluids from the combined use of cubic equations of state and solvation calculations. *Industrial & Engineering Chemistry Research, 48*(6), 3197–3205. doi:10.1021/ie801118a.

Hsieh, C.-M., & Lin, S.-T. (2009b). Prediction of 1-octanol–water partition coefficient and infinite dilution activity coefficient in water from the PR + COSMOSAC model. *Fluid Phase Equilibria, 285*(1–2), 8–14. doi:10.1016/j.fluid.2009.06.009.

Hsieh, C.-M., & Lin, S.-T. (2010). Prediction of liquid–liquid equilibrium from the Peng–Robinson+COSMOSAC equation of state. *Chemical Engineering Science, 65*(6), 1955–1963. doi:10.1016/j.ces.2009.11.036.

Justo-García, D. N., García-Sánchez, F., Díaz-Ramírez, N. L., & Díaz-Herrera, E. (2010). Modeling of three-phase vapor–liquid–liquid equilibria for a natural-gas system rich in nitrogen with the SRK and PC-SAFT EoS. *Fluid Phase Equilibria, 298*(1), 92–96. doi:10.1016/j.fluid.2010.07.012.

Lin, S.-T., & Sandler, S. I. (2002). A priori phase equilibrium prediction from a segment contribution solvation model. *Industrial & Engineering Chemistry Research, 41*(5), 899–913. doi:10.1021/ie001047w.

Lladosa, E., Montón, J. B., Burguet, M., & de la Torre, J. (2008). Isobaric (vapour + liquid + liquid) equilibrium data for (di-n-propyl ether + n-propyl alcohol + water) and (diisopropyl ether + isopropyl alcohol + water) systems at 100 kPa. *The Journal of Chemical Thermodynamics, 40*(5), 867–873. doi:10.1016/j.jct.2008.01.002.

Neoschil, J., & Chambrette, P. (1978). *Convergence Pressure Concept A Key For High Pressure Equilibria*. Society of Petroleum Engineers, Richardson,Texas, 7820.

Peng, D.-Y., & Robinson, D. B. (1976). A new two-constant equation of state. *Industrial & Engineering Chemistry Fundamentals, 15*(1), 59–64. doi:10.1021/i160057a011.

Pequenín, A., Asensi, J. C., & Gomis, V. (2010). Isobaric vapor–liquid–liquid equilibrium and vapor–liquid equilibrium for the quaternary system water–ethanol–cyclohexane–isooctane at 101.3 kPa. *Journal of Chemical & Engineering Data, 55*(3), 1227–1231. doi:10.1021/je900604a.

Prausnitz, J. M., Lichtenthaler, R. N., de Azevedo, E. G. (2004). *Molecular thermodynamics of fluid-phase equilibria* (3rd ed.). Taipei: Pearson Education Taiwan Ltd.

Rao, Y. V. C. (2003). *Chemical Engineering Thermodynamics*. Hyderabad: University Press (India) Pvt. Ltd.

Renon, H., & Prausnitz, J. M. (1968). Local compositions in thermodynamic excess functions for liquid mixtures. *AIChE Journal, 14*(1), 135–144. doi:10.1002/aic.690140124.

Sampath, V. R., & Leipziger, S. (1985). Vapor-liquid-liquid equilibria computations. *Industrial & Engineering Chemistry Process Design and Development, 24*(3), 652–658. doi:10.1021/i200030a021.

Sandler, S. I. (1999). *Chemical and Engineering Thermodynamics* (3rd ed.). New York: John Wiley & Sons.

Seader, J. D., & Henley, E. J. (2007). *Separation Process Principles* (2nd ed.). New Delhi: Wiley India (P.) Ltd.

Soave, G. (1972). Equilibrium constants from a modified Redlich-Kwong equation of state. *Chemical Engineering Science, 27*(6), 1197–1203. doi:10.1016/0009-2509(72)80096-4.

Whitson, C. H., & Michelsen, M. L. (1989). The negative flash. *Fluid Phase Equilibria, 53*, 51–71. doi:10.1016/0378-3812(89)80072-X.

Wyczesany, A. (2014). Calculation of vapor–liquid–liquid equilibria at atmospheric and high pressures. *Industrial & Engineering Chemistry Research, 53*(6), 2509–2519. doi:10.1021/ie403418p.

5

Modification in COSMO-SAC

5.1 Introduction

On the basis of the COSMO framework, Lin and Sandler (2002) derived the variation of this model to accommodate nonideality in phase equilibria calculation and predict activity coefficients. In COSMO-SAC, the activity coefficients are calculated segmentwise and the chemical potential of each molecule is determined by summing up contributions of every segment. The solvation free energy ($\Delta\underline{G}^{sol}$) was originally taken by Lin and Sandler as the summation of electrostatic contribution and van der Waals free energy. The electrostatic contribution originates due to the electrostatic interaction between solute and solvent. Van der Waals part takes cavity formation and dispersion interaction. Mathematically, it is written as

$$\Delta\underline{G}_{i/i}^{*sol} = \Delta\underline{G}_{i/i}^{*vdw} + \Delta\underline{G}_{i/i}^{*el} \tag{5.1}$$

The activity coefficient is written in terms of logarithmic summation of restoring, dispersion and cavity formation terms. In the original assumption, the dispersive term was neglected and restoring solvation free energy was taken as electrostatic contribution. The cavity formation term is defined by the Stavermann–Guggenheim combinatorial term to accommodate the free energy change of molecular size and shape differences between species.

$$\ln\left(\gamma_{i/S}\right) = \ln\left(\gamma_{i/S}^{res}\right) + \ln\left(\gamma_{i/S}^{disp}\right) + \ln\left(\gamma_{i/S}^{comb}\right) \tag{5.2}$$

$$\ln\left(\gamma_{i/S}\right) = \ln\left(\gamma_{i/S}^{res}\right) + \ln\left(\gamma_{i/S}^{comb}\right) \tag{5.3}$$

$$\ln\left(\gamma_{i/S}\right) = \frac{\Delta G_{i/S}^{*res} - \Delta G_{i/i}^{*res}}{RT} + \ln\left(\gamma_{i/S}^{SG}\right) \tag{5.4}$$

The picture depicted above does not produce the real picture of solvation of a molecule from vacuum to solvent. According to implicit continuum solvation theory, the solvation free energy is the summation of six free energies:

$$\Delta \underline{G}^{\text{sol}} = \Delta \underline{G}^{\text{el}} + \Delta \underline{G}^{\text{cav}} + \Delta \underline{G}^{\text{disp}} + \Delta \underline{G}^{\text{vib}} + \Delta \underline{G}^{\text{lib}} + \Delta \underline{G}^{\text{other}} \tag{5.5}$$

The electrostatic free energy is represented by $\Delta \underline{G}^{\text{el}}$. This originates from the polarization and electrostatic interaction among solute and solvent molecules. A cavity is needed in the solvent molecule to accept a solute molecule and the free energy change is denoted by $\Delta \underline{G}^{\text{cav}}$. Short range interactions originate from the instantaneous charges on the molecular surface, better known as dispersion free energy ($\Delta \underline{G}^{\text{disp}}$). Van der Waals free energy is the summation of dispersion and cavity terms. From the vibration, rotation and internal structure of a molecule, the last three free energies of Equation 5.5 originate. For phase equilibria calculation, these three terms are negligible, because the differences in their values between the ideal gas and the solvent are assumed to be small. With this, we can write

$$\Delta \underline{G}^{\text{sol}} = \Delta \underline{G}^{\text{el}} + \left(\Delta \underline{G}^{\text{cav}} + \Delta \underline{G}^{\text{disp}} \right) \tag{5.6}$$

Thus we obtain Equation 5.1 (we removed the subscripts for the sake of simplicity). For computation of phase equilibria, the electrostatic free energy change is calculated by the difference of interactions between the molecule in the solvent and in its fluid form. For calculating pure component properties like vapor pressure and enthalpy of vaporization, the electrostatic contribution to the free energy is represented by the sum of ideal solvation (is), charge averaging correction (cc), restoring free energy (res, including hydrogen bond corrections), that is,

$$\Delta \underline{G}_{i/i}^{*\text{el}} = \Delta \underline{G}_{i}^{*\text{is}} + \Delta \underline{G}_{i}^{*\text{cc}} + \Delta \underline{G}_{i/i}^{*\text{res}} \tag{5.7}$$

The first two terms on the right-hand side of Equation 5.7 were described in Section 3.2.5. An averaging process is used to obtain only standard segments. The standard segments have 'apparent' charge density distribution σ over an area that is larger than that of the original charge density σ^*. The COSMO-SAC model considers this standard segment surface area as one of the universal parameters. The assumption of independent segments leads to the calculation of $\Delta \underline{G}_{i/i}^{*\text{res}}$ assuming a pairwise interaction model. For that the original σ-averaging expression is slightly modified and an empirical parameter f_{decay} is introduced. The value of f_{decay} is 3.57. Other notations carry the same meaning as the previous equations. The expression is given by Equation 5.8.

$$\sigma_m = \frac{\sum_n \sigma_n^* \frac{r_n^2 r_{\text{eff}}^2}{r_n^2 + r_{\text{eff}}^2} \exp\left(-f_{\text{decay}} \frac{d_{mn}^2}{r_n^2 + r_{\text{eff}}^2}\right)}{\sum_n \frac{r_n^2 r_{\text{eff}}^2}{r_n^2 + r_{\text{eff}}^2} \exp\left(-f_{\text{decay}} \frac{d_{mn}^2}{r_n^2 + r_{\text{eff}}^2}\right)} \tag{5.8}$$

The $\Delta \underline{G}_{i/i}^{*\text{vdw}}$ is in terms of Helmholtz free energy and written as

$$\frac{\Delta \underline{G}_{i/i}^{*\text{vdw}}}{RT} = \frac{\Delta \underline{A}_{i/i}^{*\text{disp}}}{RT} + \frac{\Delta \underline{A}_{i/i}^{*\text{cav}}}{RT} - 1 \tag{5.9}$$

5.2 Hydrogen Bonding in Ionic Liquids

It has been observed in the previous chapter that the hydrogen bonding plays an important role in COSMO-SAC predictions. One of the definitions of hydrogen bonding was formulated by Pauling who stated that (Grabowski, 2011) 'under certain conditions an atom of hydrogen is attracted by rather strong forces to two atoms, instead of only one, so that it may be considered to be acting as a bond between them. This is called the hydrogen bond.' Pauling also pointed out that the hydrogen bond is situated only between most electronegative atoms and it usually interacts strongly with one of them. The later interaction is a typical covalent bond (A–H), where A is an electronegative atom like oxygen, nitrogen and fluorine and H is hydrogen. The interaction between the hydrogen and the other electronegative atom is much weaker and mostly electrostatic in nature; it is a nonbonding interaction (H···B). This system is often designated as A–H···B where B-centre (acceptor of proton) should possess at least one lone electron pair. A–H is called the proton donating bond. Pauling stated that sometimes the H···B interaction possesses characteristics of a covalent bond. In addition to the conventional H-bond, there are blueshifted H-bonds, dihydrogen A–H···H–B H-bonds, inverse H-bonds, resonance-assisted H-bonds, charge-assisted H-bonds, ionic H-bonds and C–H···H, X–H···π, π···H···π and even H···e···H H-bonds (Hunt, Ashworth, & Matthews, 2015). Thus Steiner proposed the definition of the H-bond as follows: an X–H···Y interaction is called a H-bond if it constitutes a local bond and X–H acts as a proton donor to Y (Steiner, 2002). Hydrogen bonding can be defined in terms of polarization of charges. For a strong acid, the polarization of charge will leave the donor H-atom that is electron deficient and highly positively charged. This donor H-atom will pull electron density from base and form the H-bond. A strong base will similarly push its electron density to the donor and will form the H-bond. This push and pull will depend on electronegativity, hardness/softness and polarizability of acid and base. The $[FHF]^-$ ion is an example where the proton is inserted between two negative fluorine ions, accurately in the middle of the F···F distance. Hence, both H···F interactions are equivalent. But the O–H···O and N–H···N hydrogen bonds are not equivalent because fluorine is the most electronegative atom. In case of O–H···O and N–H···N, stronger covalent bonds as well as weaker hydrogen bonds are observed. Although weaker, these hydrogen bonds contribute considerably in the physical properties like boiling point of water.

With the basic introduction of H-bonding, we concentrate our discussion on formation of hydrogen bonding in ILs. A significant overlap over these two topics has been established in recent years and predictive models are thus improved. It is possible to partition H-bonding into electrostatics, charge transfer, dispersion, polarization and exchange-repulsion terms. Thus the predictive models like COSMO-SAC are also upgraded and include structural and electronic characterization of H-bonding. Within an IL, a significantly different type of H-bonding can be formed which is unlikely for other molecular solvents. The H-bonding of IL is referred as doubly ionic hydrogen bond; primarily to differentiate from other ionic H-bonding. Based on the ability of forming H-bonding, ILs are divided into two broad classes, namely, protic and aprotic ILs. Protic ILs typically consist of Bronsted acid and Bronsted base where a proton is transferred from acid to base. Ammonium-based cations are an example of protic ILs where hydrogen is covalently bonded with cationic charge (either P or N atoms). In aprotic ILs, cations have a C–H unit attached which is the primary H-bond donor unit. Thus the indirect attachment of the H-bond donor unit makes aprotic ILs weaker H-bond forming materials than protic ILs. Despite having a tendency of forming strong H-bonding, protic ILs are less studied. Typical H-bonding formation for an ammonium-based IL is N–H···Y, whereas for aprotic ILs, the representation is C–H···Y. It is believed that proton transfer has a substantial impact on the physical and chemical properties of protic ILs. Networking is another feature of ILs due to very high density of H-bonding among ionic species. If the numbers of acceptor and donor sites are equal, that is, perfectly matched, rigid networks are formed. When ILs are mixed with other H-bonding species, additional donor/acceptors make a substantial impact on overall H-bonding. Imidazolium cations are the most widely studied IL cations. For imidazolium cations, the C–H unit is considered as the most probable donor unit. Intermediate carbene is formed by abstracting a proton from methyl and methylene groups. By increasing structural complexity, H-bonding characteristics of anions are increased. Bis-trifuoromethyl imide (Tf_2N) is considered among most complex anions in IL family because of electron-rich central N atom and pendant groups, containing O and N atoms (Hunt et al., 2015). In this chapter, we will discuss extraction of bio-oil-derived chemicals by IL where hydrogen bonding of imidazolium-[Tf_2N]-based ILs will be modelled by COSMO-SAC theory.

5.3 Gaussian-Type Probability for Hydrogen Bonding

The screening charge distribution of each segment is represented by the σ-profile which is mere the three-dimensional charge density distribution projected into a two-dimensional histogram. In original COSMO-SAC, the σ-profile is given by $p(\sigma) = A_i(\sigma)/A_i$ and for mixture,

$$p_s(\sigma) = \frac{\sum_i x_i A_i p_i(\sigma)}{\sum_i x_i A_i} \tag{5.10}$$

The screening charge density is divided into 61 equidistant bins (−0.003 e/Å2 to +0.003 e/Å^2). In modified COSMO-SAC, the bins are divided into 71 (−0.0035 e/Å^2 to +0.0035 e/Å^2). The major modification was done to incorporate hydrogen bonding in a more realistic way. There was a strict cut-off value of charge density (|0.0084| e/Å^2) to define threshold for hydrogen bonding. Thus the Hb contribution is non-zero only if one segment has a negative charge density less than zero and the other has a positive charge density greater than zero. In this way, Hb is limited to segment pairs of opposite charge and larger magnitudes. This contribution also increases directly with the charge densities of the two segments. However, the assumption that charge density alone determines Hb behavior is not fully consistent with experimental observations, which suggest that, for the most part, Hb acceptors are limited to small electronegative atoms (e.g. nitrogen, oxygen and fluorine) and that Hb donors are limited to hydrogen atoms bound to such acceptors. The use of a strict cut-off for Hb behavior is also not physically meaningful. These weaknesses leave it prone to occasional error.

Lin, Chang, Wang, Goddard and Sandler (2004) first proposed the modifications in COSMO-SAC and predicted vapor pressure and enthalpy of vaporization. It includes the van der Waals interactions in solvation free energy calculations and also the ideal solvation term and charge averaging correction term in electrostatic contribution. The residual term originally includes misfit and hydrogen bonding energies. They also proposed to use two different σ-profiles: one for nonhydrogen bonding part (non-Hb part) and another for hydrogen bonding part (Hb part). The concept of separate σ-profiles for hydrogen bonding and nonhydrogen bonding segments is also used in mixtures when Wang, Sandler and Chen (2007) did modifications in original COSMO-SAC. On the basis of the definitions of hydrogen bonding, they classified molecules into three categories described as follows:

1. Compounds that have approximately neutral segments; n-butane is an example of this type of molecule.

2. Compounds that have neutral segments for the σ-profile of the nonhydrogen bonding part and hydrogen bonding acceptor segments from among the O, N, F atoms that form the Hb σ-profile. Nitromethane is an example of this type of molecule.
3. Compounds that not only have neutral segments for the nonhydrogen bonding σ-profile but also have both hydrogen bonding acceptor segments (from the O, N, F atoms) and hydrogen bonding donor segments (from H atoms connected to the O, N, F atoms); both contribute to the hydrogen bonding σ-profile. Water is an example of this third type of molecule.

There are three ways to account for hydrogen bonding in the Hb part of the σ-profile (Wang et al., 2007).

1. The simplest case is that acceptor and donor segments in the Hb σ-profile are considered to be sufficiently polar to result in hydrogen bonding, regardless of the magnitude of charge density.
2. The second case is that a separate σ-profile is considered for hydrogen bonding only when their charge density exceeds a certain threshold value (e.g. $|0.0084|$ e/Å^2), which leads to the step function change in probability, which was used in the original COSMO-SAC model.
3. The third case is that the probability of forming a hydrogen bond is based on a continuous probability distribution function of charge density so that for the acceptor and donor segments, higher the charge density, the greater the possibility of forming a hydrogen bond. In this procedure, the Hb σ-profile is weighted with the probability density function and the difference between the Hb and weighted Hb profile is added back into the non-Hb σ-profile.

The expression for the Gaussian-type probability ($P^{Hb}(\sigma)$) is given by Equation 5.11.

$$P^{Hb}(\sigma) = 1 - \exp\left(-\frac{\sigma^2}{2\sigma_0^2}\right) \tag{5.11}$$

where $\sigma_0 = 0.007\ \mathrm{e}/\text{Å}^2$ and σ values are calculated from Equation 5.8. Figure 5.1 represents how $P^{Hb}(\sigma)$ changes with σ.

Each bin in the Hb σ-profile is multiplied (reduced) by its $P^{Hb}(\sigma)$ value and the portion of the Hb σ-profile thus removed is added back into the corresponding bin of the non-Hb σ-profile. In this way, the overall σ-profile is conserved. From a physical standpoint, this approach allows the model to limit Hb interactions to a certain fraction of the available Hb donor and

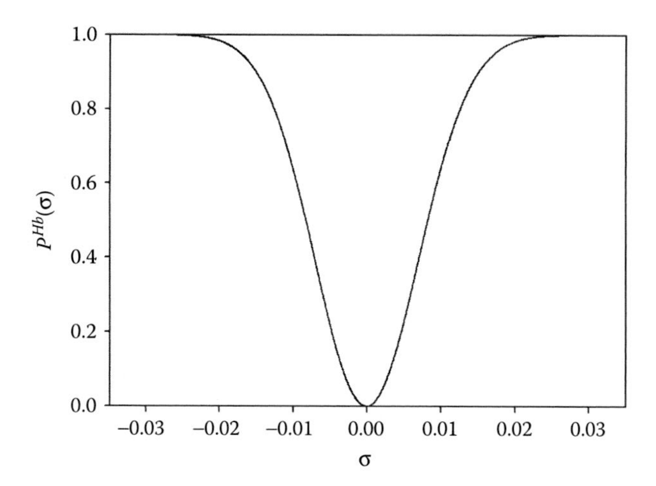

FIGURE 5.1
$P^{Hb}(\sigma)$ changes with σ (e/Å²).

acceptor segments. The greater the magnitude of σ for a given segment, the more likely that segment is to exhibit Hb. This provides a physically reasonable alternative to merely limiting Hb to segments whose σ value exceeds an arbitrary threshold. With this explanation the expressions for the Hb σ-profile and non-Hb σ-profile will be (Equations 5.14 through 5.16, respectively)

$$p_i^{non\text{-}Hb}(\sigma) = \frac{n_i^{non\text{-}Hb}(\sigma)}{n_i^{non\text{-}Hb}} = \frac{A_i^{non\text{-}Hb}(\sigma)}{A_i^{non\text{-}Hb}} \tag{5.12}$$

$$p_i^{Hb}(\sigma) = \frac{n_i^{Hb}(\sigma)}{n_i^{Hb}} = \frac{A_i^{Hb}(\sigma)}{A_i^{Hb}} \tag{5.13}$$

where $n_i = A_i / a_{\text{eff}}$,

$$p_{\text{new}}^{Hb}(\sigma) = p^{Hb}(\sigma) p_i^{Hb}(\sigma) = \frac{A_i^{Hb}(\sigma)}{A_i^{Hb}} \tag{5.14}$$

$$p_{\text{new}}^{non\text{-}Hb}(\sigma) = p_i^{non\text{-}Hb}(\sigma) + p^{Hb}(\sigma)\left[1 - p_i^{Hb}(\sigma)\right] = \frac{A_i^{non\text{-}Hb}(\sigma)}{A_i^{non\text{-}Hb}} + \frac{A_i^{Hb}(\sigma)}{A_i^{Hb}}\left[1 - p_i^{Hb}(\sigma)\right] \tag{5.15}$$

$$p_i(\sigma) = p_{\text{new}}^{Hb}(\sigma) + p_{\text{new}}^{non\text{-}Hb}(\sigma) \tag{5.16}$$

where i is the compound considered. $A_i^{Hb}(\sigma)$ is the summation of hydrogen bonding segment areas, $A_i^{non\text{-}Hb}(\sigma)$ is the summation of nonhydrogen bonding areas and A_i is the total segment areas. The interaction energy is the summation of misfit energy and hydrogen bonding energy. The concept of a separate σ-profile was later rigorously investigated for the atom-specific (O, N, F) hydrogen bonding σ-profile. We will cover most of it in the subsequent sections. For now, with the hydrogen bonding and nonhydrogen bonding σ-profile, the expression for segment activity coefficients is also modified. The interaction energy is the summation of misfit energy and hydrogen bonding energy. Recalling the original expression for interaction energy,

$$\Delta W(\sigma_m, \sigma_n) = E_{mf} + E_{Hb} \tag{5.17}$$

This equation is rewritten as

$$\Delta W(\sigma_m^t, \sigma_n^s) = c_{es}(\sigma_m^t + \sigma_n^s)^2 - c_{Hb}(\sigma_m^t, \sigma_n^s)(\sigma_m^t - \sigma_n^s)^2 \tag{5.18}$$

where s and t can be *Hb* (hydrogen bonding) or *non-Hb* (nonhydrogen bonding). σ_m and σ_n are the average charge densities. c_{es} is given by

$$c_{\text{es}} = f_{\text{pol}}\left(\frac{0.3a_{\text{eff}}^{3/2}}{2\varepsilon_0}\right) \tag{5.19}$$

where $f_{\text{pol}} = 0.6916$ is the polarization factor and ε_0 is the permittivity of vacuum. $c_{Hb}(\sigma_m^t, \sigma_n^s)$ is given by

$$c_{Hb}(\sigma_m^t, \sigma_n^s) = \begin{cases} c_{Hb} & s = t = Hb, \quad \sigma_m^t \times \sigma_n^s < 0 \\ 0 & \text{otherwise} \end{cases} \tag{5.20}$$

The segment activity coefficient (Γ) of segment m with the charge density of σ_m is determined from the σ-profile and is given by Equation 5.21, which is solved self-consistently.

$$\ln\left(\Gamma_i^t(\sigma_m^t)\right) = -\ln\left[\sum_s^{non\text{-}Hb,Hb} \sum_{\sigma_n} p_i^s(\sigma_n^s)\Gamma_i^s(\sigma_n^s)\exp\left(-\frac{\Delta W(\sigma_m^t, \sigma_n^s)}{RT}\right)\right] \tag{5.21}$$

$$\ln\left(\Gamma_S^t(\sigma_m^t)\right) = -\ln\left[\sum_s^{non\text{-}Hb,Hb} \sum_{\sigma_n} p_S^s(\sigma_n^s)\Gamma_S^s(\sigma_n^s)\exp\left(-\frac{\Delta W(\sigma_m^t, \sigma_n^s)}{RT}\right)\right]$$

Finally, the restoring free energy is obtained by Equation 5.22.

$$\frac{\Delta \underline{G}_{i/j}^{*\text{res}}}{RT} = \frac{A_i}{a_{\text{eff}}} \sum_{s}^{non\text{-}Hb,Hb} \sum_{\sigma_m} p_i^s\left(\sigma_m^s\right) \ln \Gamma_j^s\left(\sigma_m^s\right) \tag{5.22}$$

For solvent j will be replaced by sol and for pure compound it will be i. So the activity coefficient of a compound will be given by

$$\ln\left(\gamma_{i/\text{sol}}\right) = \frac{\Delta \underline{G}_{i/\text{sol}}^{*\text{res}} - \Delta \underline{G}_{i/i}^{*\text{res}}}{RT} + \frac{\Delta \underline{G}_{i/\text{sol}}^{*\text{disp}} - \Delta \underline{G}_{i/i}^{*\text{disp}}}{RT} + \frac{\Delta \underline{G}_{i/\text{sol}}^{*\text{cav}} - \Delta \underline{G}_{i/i}^{*\text{cav}}}{RT} \tag{5.23}$$

As already discussed, cavity formation free energy is modelled by the Staverman–Guggenheim combinatorial term. So the above equation reduces to

$$\ln\left(\gamma_{i/\text{sol}}\right) = \frac{\Delta \underline{G}_{i/\text{sol}}^{*\text{res}} - \Delta \underline{G}_{i/i}^{*\text{res}}}{RT} + \frac{\Delta \underline{G}_{i/\text{sol}}^{*\text{disp}} - \Delta \underline{G}_{i/i}^{*\text{disp}}}{RT} + \ln\left(\gamma_{i/\text{sol}}^{\text{SG}}\right) \tag{5.24}$$

The restoring free energy makes dominant contribution in calculating the activity coefficient model especially for polar systems. The dispersion term, which is important for pure property calculations, has a very small contribution in the activity coefficient calculation, as a result of its cancellation between the solvent and pure solute phases. So, Equation 5.24 reduces to

$$\ln\left(\gamma_{i/\text{sol}}\right) = \frac{\Delta \underline{G}_{i/\text{sol}}^{*\text{res}} - \Delta \underline{G}_{i/i}^{*\text{res}}}{RT} + \ln\left(\gamma_{i/\text{sol}}^{\text{SG}}\right) \tag{5.25}$$

$$\begin{aligned} \ln\left(\gamma_{i/S}\right) &= \frac{\Delta G_{i/S}^{*\text{sol}} - \Delta G_{i/i}^{*\text{sol}}}{RT} + \ln\left(\gamma_{i/S}^{\text{SG}}\right) \\ &= n \sum_{s}^{non\text{-}Hb,Hb} \sum_{\sigma_m} p\left(\sigma_m^s\right)\left[\ln\left(\Gamma_{i/S}^s\left(\sigma_m^s\right)\right) - \ln\left(\Gamma_{i/i}^s\left(\sigma_m^s\right)\right)\right] + \ln\left(\gamma_{i/S}^{\text{SG}}\right) \end{aligned} \tag{5.26}$$

The temperature dependence of the coefficient of electrostatic interactions is incorporated by Hsieh, Sandler and Lin (2010). This is particularly interesting for the nonhydrogen bonding σ-profile and thus electrostatic interactions vary with temperature.

$$c_{\text{es}} = A_{\text{es}} + \frac{B_{\text{es}}}{T^2} \tag{5.27}$$

where c_{es} is the temperature-dependent coefficient of misfit energy and A_{es} and B_{es} are two constants whose values are obtained regressing experimental vapor-liquid equilibria (VLE) data. In Equation 5.27, T is temperature in K (Table 5.1).

TABLE 5.1

Values of A_{es} and B_{es}

Parameter	Value	Unit
A_{es}	6525.69	(kcal/mol)(Å^4/e^2)
B_{es}	1.4859×10^8	(kcal/mol)(Å^4/e^2)

Source: Hsieh, C.-M. et al., *Fluid Phase Equilib.*, 297, 90–97, 2010.

The concept of a separate σ-profile later subdivided into functional group-specific *Hb* σ-profiles. The idea became popular because the strengths of hydrogen bonds differ according to the environments of atoms involved. Hsieh et al. (2010) first differentiated surfaces of hydroxyl groups ($A_i^{OH}(\sigma)$) from other hydrogen bonding surfaces ($A_i^{OT}(\sigma)$).

$$p(\sigma) = p^{non\text{-}Hb}(\sigma) + p^{OH}(\sigma) + p^{OT}(\sigma) \tag{5.28}$$

$$p^{OH}(\sigma) = p_0^{OH}(\sigma)p^{Hb}(\sigma) \tag{5.29}$$

$$p^{OT}(\sigma) = p_0^{OT}(\sigma)p^{Hb}(\sigma) \tag{5.30}$$

$$p^{non\text{-}Hb}(\sigma) = p_0^{non\text{-}Hb}(\sigma) + \left[p_0^{OH}(\sigma) + p_0^{OT}(\sigma)\right]\left[1 - p^{Hb}(\sigma)\right] \tag{5.31}$$

where '*T*' represents N or F atoms.

$$c_{Hb}\left(\sigma_m^t, \sigma_n^s\right) = \begin{cases} c_{OH\text{-}OH} & s = t = OH, \quad \sigma_m^t \times \sigma_n^s < 0 \quad c_{OH\text{-}OH} = 4013.78 \\ c_{OT\text{-}OT} & s = t = OT, \quad \sigma_m^t \times \sigma_n^s < 0 \quad c_{OT\text{-}OT} = 932.31 \\ c_{OH\text{-}OT} & s = OH,\ t = OH, \quad \sigma_m^t \times \sigma_n^s < 0 \quad c_{OH\text{-}OT} = 3016.43 \\ 0 & \text{otherwise} \end{cases}$$

The values of c_{Hb} are in (kcal/mol)(Å^4/e^2) and are taken from the literature (Hsieh et al., 2010).

Like oxygen, a separate *Hb* σ-profile for nitrogen containing compounds was later proposed. The primary concern is amino groups where primary, secondary and tertiary amines are possible. All the *Hb* acceptors have lone pairs (nitrogen has one, oxygen has two and fluorine has three). These lone pairs show high directional specificity of hydrogen bonds and alignment with donor determines the strength for interaction. Thus the σ-profile becomes

$$p(\sigma) = p^{non\text{-}Hb}(\sigma) + p^{\text{hydro}}(\sigma) + p^{\text{amino}}(\sigma) + p^{\text{other}}(\sigma) \tag{5.32}$$

$$c_{Hb}\left(\sigma_m^t,\sigma_n^s\right)=\begin{cases} c_{HH} & \text{if } s=t=\text{hydro and} & \sigma_m^t\times\sigma_n^s<0 & c_{HH}=1757.9468\\ c_{AA} & \text{if } s=t=\text{amino and} & \sigma_m^t\times\sigma_n^s<0 & c_{AA}=1121.4047\\ c_{OO} & \text{if } s=t=\text{other and} & \sigma_m^t\times\sigma_n^s<0 & c_{OO}=1757.9468\\ c_{HA} & \text{if } s=\text{hydro}, t=\text{amino and} & \sigma_m^t\times\sigma_n^s<0 & c_{HA}=2462.3206\\ c_{HO} & \text{if } s=\text{hydro}, t=\text{other and} & \sigma_m^t\times\sigma_n^s<0 & c_{HO}=933.4108\\ c_{AO} & \text{if } s=\text{amino}, t=\text{other and} & \sigma_m^t\times\sigma_n^s<0 & c_{AO}=2057.9712\\ 0 & \text{otherwise} \end{cases}$$

The values of c_{Hb} are in (kcal/mol)(Å^4/e^2) and are taken from the literature (Hsieh & Lin, 2012).

5.3.1 Computational Details

In modified COSMO-SAC, two separate σ-profiles have to be generated using COSMO file information more efficiently. The first step is identification of electronegative and hydrogen atoms. In the Gaussian03 COSMO output file, elements are identified by their respective atomic numbers (1 for hydrogen, 6 for carbon, 7 for nitrogen, 8 for oxygen, etc.). As carbon does not participate in hydrogen bonding, it is not taken into the matrix and the rest of the elements along with their respective positions (1, 2, 3, etc.) are stored in a matrix (matrix A). This kind of storing will be useful in identifying the segments of a particular atom.

Next, all hydrogen atoms do not participate in hydrogen bonding. As for hydrocarbon systems, the hydrogen that is bonded with carbon will not participate in hydrogen bonding. Based on geometry optimization, performed in Quantum Chemistry packages like Gaussian03, the C–H bond length (e.g. for octane optimized by HF/6-31G*, bond length varies from 1.085722 Å to 1.088926 Å) is chosen. So, if the distance between a carbon and hydrogen comes within this bond length, the hydrogen is bonded with the carbon. So hydrogen is removed from matrix A. Now the matrix will constitute only the elements that are capable of hydrogen bonding. But all segments of these elements will not participate in hydrogen bonding. Positively charged segments from the surfaces of all nitrogen, oxygen and fluorine atoms are considered as Hb acceptors. Negatively charged segments from the surfaces of hydrogen atoms bound to any one of the three acceptor atoms, that is, nitrogen, oxygen and fluorine, are considered Hb donors. When those are identified, the average charge densities (calculated by Equation 5.8) and segment areas are separated. The segment areas are then discretized in 71 bins with their respective charge densities and σ-profiles for Hb and non-Hb formed (Equations 5.14 and 5.15 are used). The interaction energy matrix is also modified because the hydrogen bonding interaction will exist only when one hydrogen bonding (Hb) discrete bin would interact with another Hb discrete bin; otherwise only misfit energy will be present. Note that, for the case of interaction energy calculation, average charge densities are replaced with the 71 discrete charge densities. This allows using the σ-profiles (both Hb and non-Hb) in Equations 5.21 and 5.22. As proposed, two separate σ-profiles (Hb and non-Hb) would be generated for each compound and σ-profiles would be the summation of these. Alcohols and amines have Hb acceptor and donor segments and non-Hb segments, whereas ketones and ethers do not have any Hb donor segments. According to this concept only H, O, N, F will participate in hydrogen bonding. Chlorine (Cl) is not included in this list; so, chloroform will have no hydrogen bonding segments. These separate σ-profiles are shown in Figure 5.2.

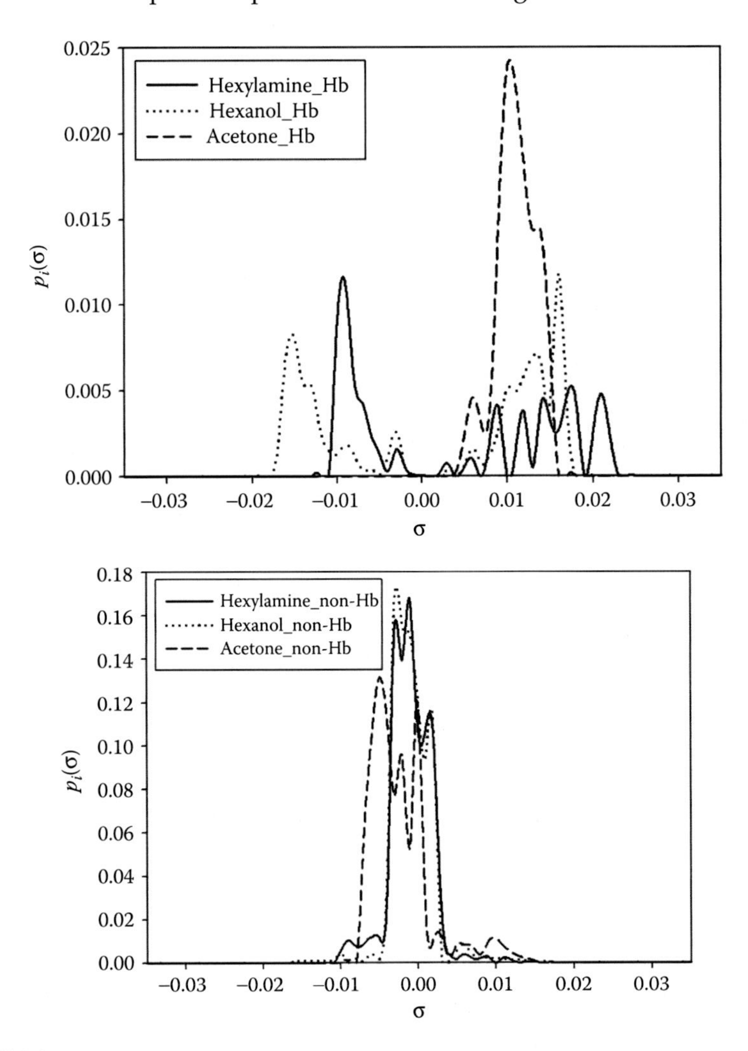

FIGURE 5.2
Hb and non-Hb σ-profiles of hexylamine, hexanol, acetone and chloroform (σ in e/Å²).

(Continued)

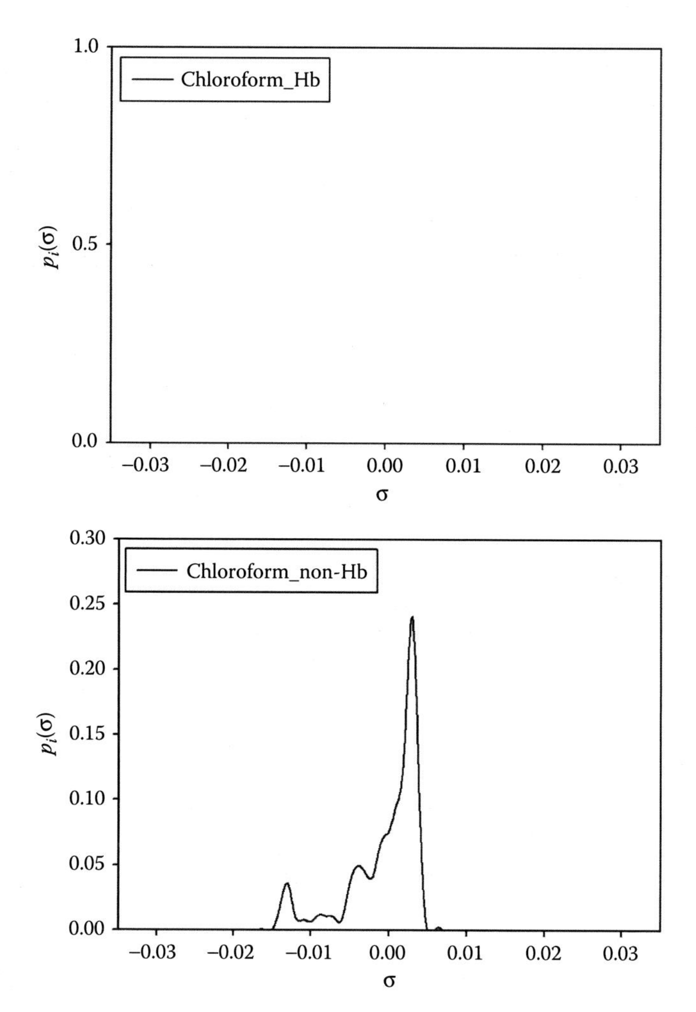

FIGURE 5.2 (Continued)
Hb and non-Hb σ-profiles of hexylamine, hexanol, acetone and chloroform (σ in e/Å^2).

5.4 Application of Modified COSMO-SAC in LLE*

One of the applications where the hydrogen bonding is put to test can be aqueous-based systems. On a similar line, lignocellulosic biomass has received a great attention as a renewable energy resource as they can improve energy

* Section 5.4 reprinted (adapted) from A. Bharti, T. Banerjee, Enhancement of bio-oil-derived chemicals in aqueous phase using ionic liquids: experimental and COSMO-SAC predictions using a modified hydrogen bonding expression. *Fluid Phase Equilibria*. 400, 27–37, 2015. Copyright 2015, with permission from Elsevier.

security and reduce carbon emissions (Naik, Goud, Rout, & Dalai, 2010; Sims, Mabee, Saddler, & Taylor, 2010). The last decade saw considerable research in the fast pyrolysis process for the production of liquid fuel and chemicals from biomass. Fast pyrolysis of biomass produces 60–75 wt% of liquid bio-oil, 15–25 wt% of solid char and 10–20 wt% of noncondensable gases depending on the feedstock used (Bridgwater, 2003, 2012). Several chemicals have been identified in bio-oil of which the most abundant and of interest are glycolaldehyde (0.9–13 wt%), acetic acid (0.5–12 wt%), formic acid (0.3–9.1 wt%), acetol (0.7–7.4 wt%), furfural alcohol (0.1–5.2 wt%) and furfural (0.1–1.1 wt%) (Bharti & Banerjee, 2015). Due to the high concentration of the value-added chemical compounds, production of chemicals from bio-oil has received considerable interest. To extract these chemicals from bio-oil, LLE is a popular technique and a significant portion of diluents were found in the raffinate phase or the aqueous-rich phase of LLE. This necessitates the use of novel solvents which can serve two purposes, namely negligible concentration in the aqueous phase and higher selectivity and distribution for bio-oil-derived chemicals. Acetic acid and furfural are chosen as model compounds to extract from the aqueous phase of bio oil and the phase behavior was predicted by the modified COSMO-SAC approach. Based on preliminary LLE experiments, hydrophobic ionic liquids are used for the extraction of acetic acid and furfural from aqueous solution. Hydrophobic characteristic of ILs mainly depend on the nature of the anion. Thus commercial hydrophobic imidazolium-based ionic liquid, 1-butyl-3-methylimidazolium bis(trifluoromethylsulfonyl) imide ([BMIM][Tf_2N]), was investigated for the extraction at $T = 298.15$ K and atmospheric pressure.

A conformal study was carried out to have a close look into the effect of hydrogen bonding in the systems. Furfural (C_4H_3OCHO, FUR) has two planar rotational conformers: *cis*-OO and *trans*-OO (Rivelino, Coutinho, & Canuto, 2002; Rogojerov, Keresztury, & Jordanov, 2005). These two conformers are formed by the rotation of the carbonyl (–CHO) group around the C–C single bond. In *cis*-OO, carbonyl O and furfural ring O atom remain on the same side of the C–C single bond, whereas in *trans*-OO, they remain opposite to each other with respect to the C–C bond. Figure 5.3a and b shows the optimized equilibrium geometries of two conformers with important geometrical parameters and total energies. The two conformers can be best described by the dihedral angle O1-C4-C5-O2. The calculated dihedral angles for conformers are *cis*-OO (0°) and *trans*-OO (180°). Based on the optimized geometries, the energy order of the conformers is predicted to be *trans*-OO (–342.5186444 Hartree) < *cis*-OO (–342.5179797 Hartree). This indicates that the *trans*-OO conformer is more stable than *cis*-OO by 1.75 kJ/mol. Thus the *trans*-OO conformer structure is then used for COSMO file generation. In a similar manner, acetic acid monomer (CH_3COOH, AA) has two conformers: *trans*-AA and *cis*-AA. In *trans*-AA, hydroxyl H points along the carbonyl O, whereas in *cis*-AA, it points opposite to carbonyl O (Maçôas, Khriachtchev, Fausto, & Räsänen, 2004; Senent, 2001). The equilibrium geometries of two conformers along with important geometrical parameters and total energies are shown in Figure 5.3c and d. The *trans*-AA has been found to be more stable than the *cis*-AA by 25.3 kJ/mol due to formation of the hydrogen bond between carbonyl O and hydroxyl H with the H-bond length (HbL) of 2.279 Å. Here the *trans*-AA conformer structure is then used for COSMO file generation. In the case of ionic liquid, the COSMO file is generated separately for the cation and anion. Based on the COSMO file, the sigma profiles for the molecules are calculated by the modified COSMO-SAC approach as

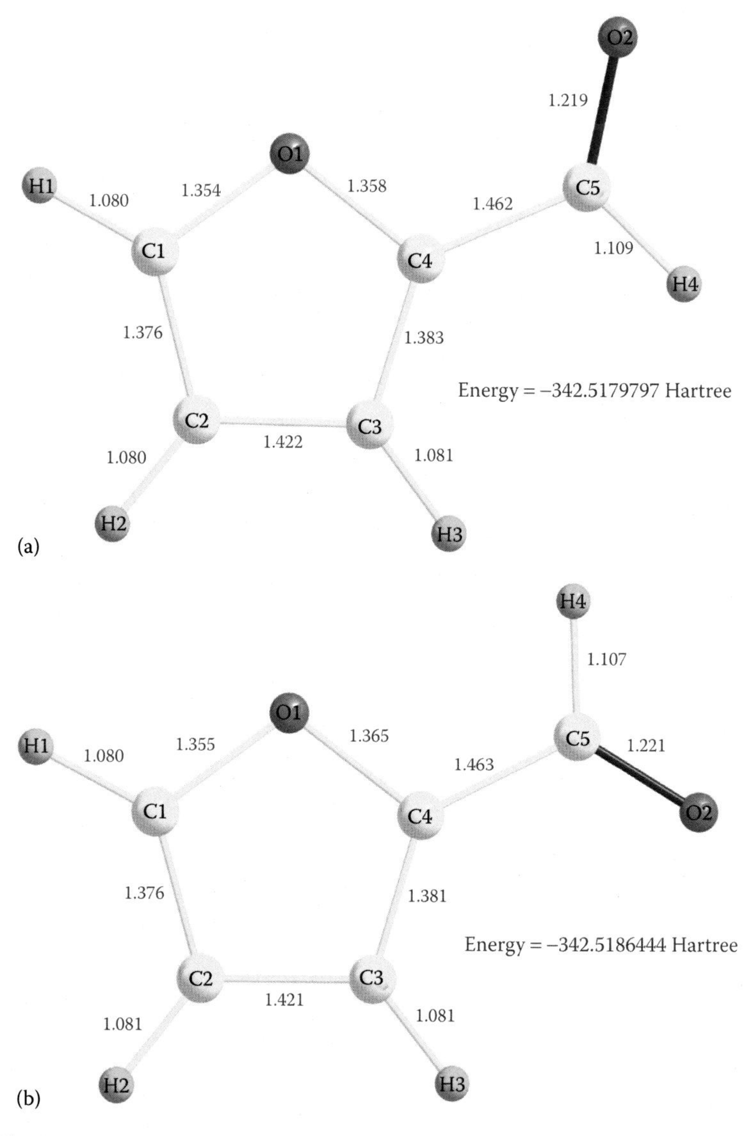

FIGURE 5.3
Optimized equilibrium geometry of (a) *cis*-OO furfural and (b) *trans*-OO furfural. (*Continued*)

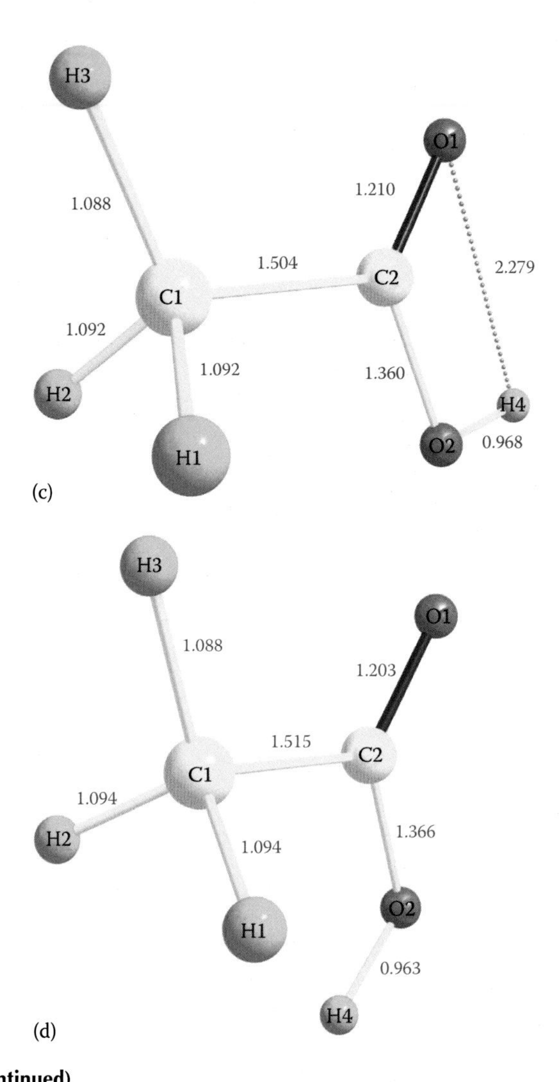

FIGURE 5.3 (Continued)
Optimized equilibrium geometry of (c) *trans*-AA (total energy = –228.5682807 Hartree) and (d) *cis*-AA (total energy = –228.5586313 Hartree).

explained in Section 5.3. For this study, we only apply the modification from Equations 5.8 through 5.26.

Thus the adoption of a continuous probability distribution function of charge density for the acceptor and donor segments increases the possibility of forming a hydrogen bond (Hb). All the compounds considered in this case study, namely, [BMIM][Tf_2N], water, furfural and acetic acid, not only have neutral segments for the nonhydrogen bonding σ-profile but also have both hydrogen bonding acceptor segments (from the O, N, F atoms) and hydrogen bonding donor segments (from H atoms connected to the O, N, F atoms);

both contribute to the hydrogen bonding σ-profile. It should be noted that the parameters a_{eff} and c_{Hb} were not reoptimized in this work. Thus the hydrogen bonding portion of the sigma profile was obtained from the combination of the electronegative atom and hydrogen atom only. The hydrogen bonding, nonhydrogen bonding and the total sigma profile for all the species namely BMIM, Tf_2N, furfural and acetic acid are depicted in Figures 5.4 through 5.6.

As depicted, the COSMO-SAC model reproduces the correct raffinate phase composition for both the systems. It also predicts a negligible presence of IL in the raffinate phase which is the same as of the experimental trend. The most important aspect of this study is the correct prediction of the slopes of the experimental and predicted tie lines. It should be noted that there was no modification of the COSMO-SAC parameters, where the model

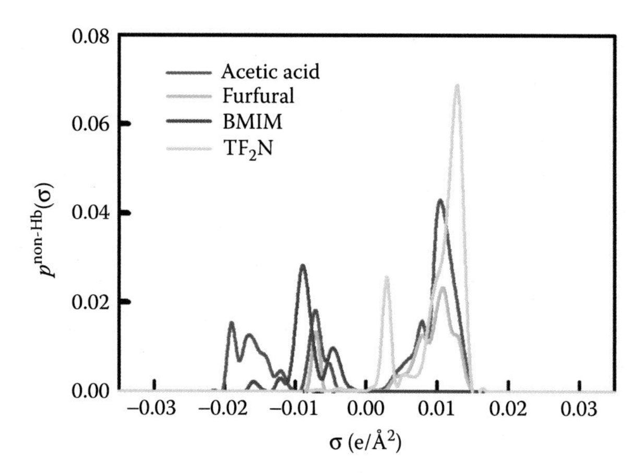

FIGURE 5.4
Hydrogen bonding sigma profile for all components.

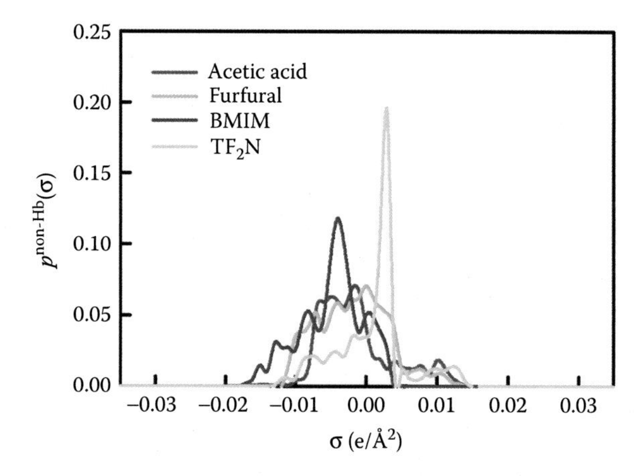

FIGURE 5.5
Nonhydrogen bonding sigma profile for all components.

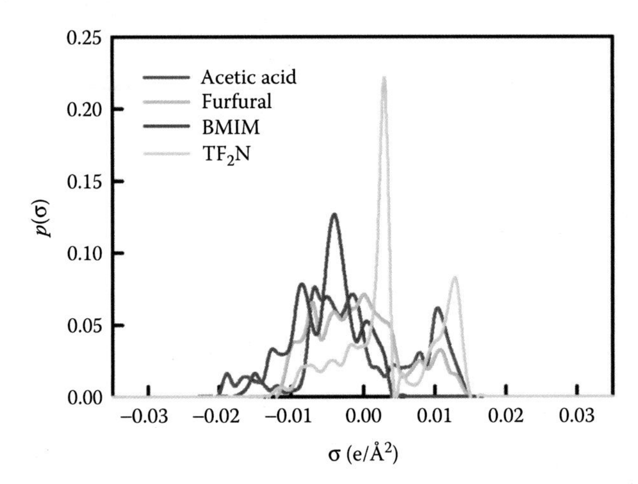

FIGURE 5.6
Total sigma profile (*Hb* + *non-Hb*) for all components.

was reimplemented from Wang et al. (2007). This further verifies the NMR peak assignments and the experimental procedure. The goodness of fit is measured by the root-mean-square deviation (RMSD), which provides the RMSD values of 2.9% (acetic acid) and 2.2% (furfural). These RMSD values indicate a good degree of consistency of the experimental LLE data for the studied systems at 298.15 K (Figures 5.7 and 5.8).

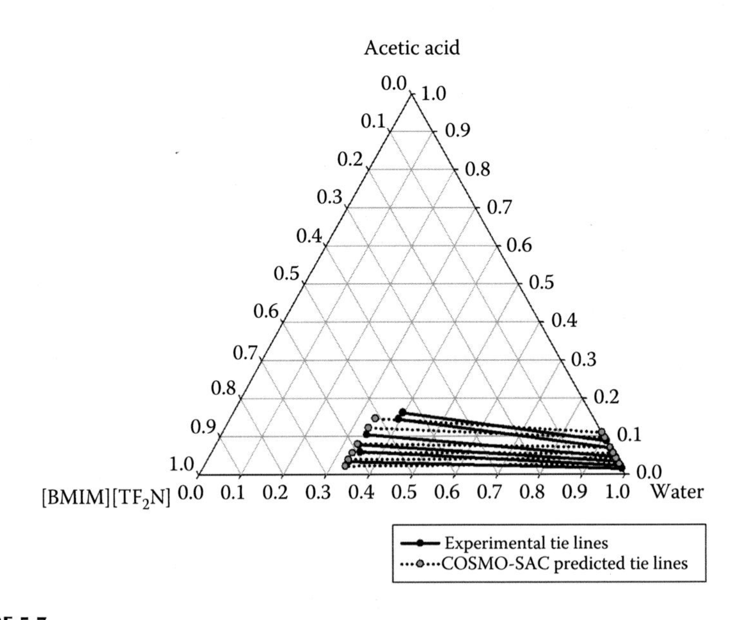

FIGURE 5.7
Experimental and COSMO-SAC predicted tie lines for the ternary system: [BMIM][TF_2N]–acetic acid–water at $T = 298.15$ K and $p = 1$ atm.

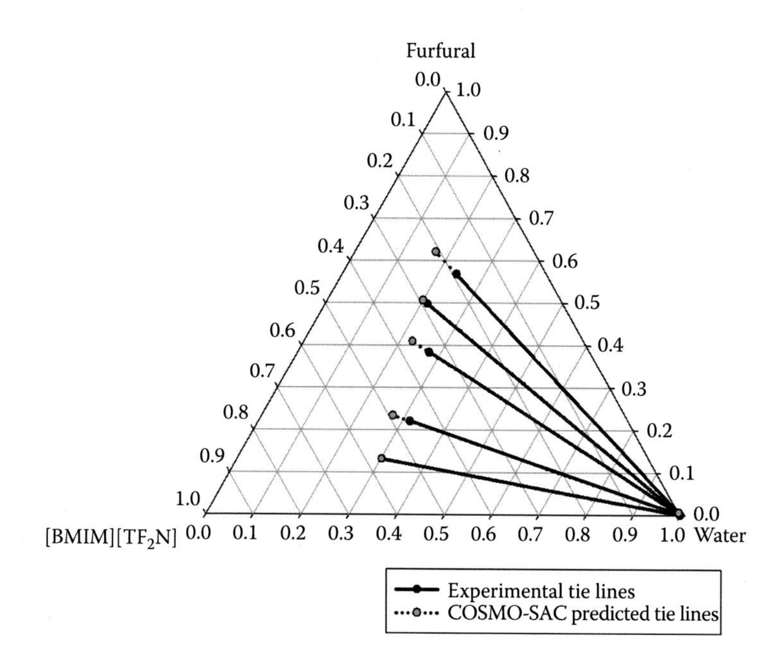

FIGURE 5.8
Experimental and COSMO-SAC predicted tie lines for the ternary system: [BMIM][TF$_2$N]–furfural–water at $T = 298.15$ K and $p = 1$ atm.

References

Bharti, A., & Banerjee, T. (2015). Enhancement of bio-oil derived chemicals in aqueous phase using ionic liquids: Experimental and COSMO-SAC predictions using a modified hydrogen bonding expression. *Fluid Phase Equilibria, 400*, 27–37. doi:10.1016/j.fluid.2015.04.029.

Bridgwater, A. V. (2003). Renewable fuels and chemicals by thermal processing of biomass. *Chemical Engineering Journal, 91*(2–3), 87–102. doi:10.1016/S1385-8947(02)00142-0.

Bridgwater, A. V. (2012). Review of fast pyrolysis of biomass and product upgrading. *Biomass and Bioenergy, 38*, 68–94. doi:10.1016/j.biombioe.2011.01.048.

Grabowski, S. J. (2011). What is the covalency of hydrogen bonding? *Chemical Reviews, 111*(4), 2597–2625. doi:10.1021/cr800346f.

Hsieh, C.-M., & Lin, S.-T. (2012). First-principles prediction of phase equilibria using the PR + COSMOSAC equation of state. *Asia-Pacific Journal of Chemical Engineering, 7*, S1–S10. doi:10.1002/apj.608.

Hsieh, C.-M., Sandler, S. I., & Lin, S.-T. (2010). Improvements of COSMO-SAC for vapor–liquid and liquid–liquid equilibrium predictions. *Fluid Phase Equilibria, 297*(1), 90–97. doi:10.1016/j.fluid.2010.06.011.

Hunt, P. A., Ashworth, C. R., & Matthews, R. P. (2015). Hydrogen bonding in ionic liquids. *Chemical Society Reviews, 44*(5), 1257–1288. doi:10.1039/C4CS00278D.

Lin, S.-T., Chang, J., Wang, S., Goddard, W. A., & Sandler, S. I. (2004). Prediction of vapor pressures and enthalpies of vaporization using a COSMO solvation model. *The Journal of Physical Chemistry A, 108*(36), 7429–7439. doi:10.1021/jp048813n.

Lin, S.-T., & Sandler, S. I. (2002). A priori phase equilibrium prediction from a segment contribution solvation model. *Industrial & Engineering Chemistry Research, 41*(5), 899–913. doi:10.1021/ie001047w.

Maçôas, E. M. S., Khriachtchev, L., Fausto, R., & Räsänen, M. (2004). Photochemistry and vibrational spectroscopy of the trans and cis conformers of acetic acid in solid Ar. *The Journal of Physical Chemistry A, 108*(16), 3380–3389. doi:10.1021/jp037840v.

Naik, S. N., Goud, V. V., Rout, P. K., & Dalai, A. K. (2010). Production of first and second generation biofuels: A comprehensive review. *Renewable and Sustainable Energy Reviews, 14*(2), 578–597. doi:10.1016/j.rser.2009.10.003.

Rivelino, R., Coutinho, K., & Canuto, S. (2002). A Monte Carlo-quantum mechanics study of the solvent-induced spectral shift and the specific role of hydrogen bonds in the conformational equilibrium of furfural in water. *The Journal of Physical Chemistry B, 106*(47), 12317–12322. doi:10.1021/jp026318q.

Rogojerov, M., Keresztury, G., & Jordanov, B. (2005). Vibrational spectra of partially oriented molecules having two conformers in nematic and isotropic solutions: furfural and 2-chlorobenzaldehyde. *Spectrochimica Acta Part A: Molecular and Biomolecular Spectroscopy, 61*(7), 1661–1670. doi:10.1016/j.saa.2004.11.043.

Senent, M. L. (2001). Ab initio determination of the torsional spectra of acetic acid. *Molecular Physics, 99*(15), 1311–1321. doi:10.1080/00268970110048374.

Sims, R. E. H., Mabee, W., Saddler, J. N., & Taylor, M. (2010). An overview of second generation biofuel technologies. *Bioresource Technology, 101*(6), 1570–1580. doi:10.1016/j.biortech.2009.11.046.

Steiner, T. (2002). The hydrogen bond in the solid state. *Angewandte Chemie International Edition, 41*(1), 48–76. doi:10.1002/1521-3773(20020104)41:1<48::AID-ANIE48>3.0.CO;2-U.

Wang, S., Sandler, S. I., & Chen, C.-C. (2007). Refinement of COSMO–SAC and the applications. *Industrial & Engineering Chemistry Research, 46*(22), 7275–7288. doi:10.1021/ie070465z.

6

Particle Swarm Optimization and Application to Liquid–Liquid Equilibrium

6.1 Introduction

Low volatile phosphonium ILs have proved to be better solvents as compared to volatile organic solvents from our liquid–liquid equilibrium (LLE) experiments in Chapter 2. However, the experiments were carried out on the laboratory scale. The separation has not been implemented on an industrial scale to date. For transforming laboratory data for an industrial application, a process optimization study is necessary. In the past, many popular stochastic algorithms such as Genetic Algorithm (GA) (Goldberg, 1989), Simulated Annealing (SA) (Kirkpatrick, Gelatt, & Vecchi, 1983), particle swarm optimization (PSO) (Eberhart & Kennedy, 1995a, 1995b), Ant colony optimization (ACO) (Colorni, Dorigo, & Maniezzo, 1991; Dorigo, 1992), Differential Evolution (DE) (Storn & Price, 1997) and Self-Organising Migrating Algorithm (SOMA) (Zelinka, 2004) have been investigated for optimization in science and engineering.

PSO is an evolutionary algorithm based on social behavior of birds in swarm. The initial position and velocity of each particle are initiated randomly. During simulation, each particle in swarm (population) updates its position and velocity based on its experience as well as neighbors' experience within the search space. PSO is robust as it evaluates fewer function values during simulation than GA (Hassan, Cohanim, & de Weck, 2005; Sivanandam & Deepa, 2009). Ethni, Zahawi, Giaouris and Acarnley (2009) showed that PSO shows more success rate to reach the target optimum value than SA. PSO is also more preferable than ACO due to high success rate and solution quality (Elbeltagi, Hegazy, & Grierson, 2005). Keeping the shortcomings of other methodologies in mind, we have chosen the PSO technique for optimizing the flow rate and number of stages in a multi-stage extractor.

A multi-stage extractor containing more than two components requires detailed design like temperature, pressure, flow rate and composition in each stage. These are achieved by solving material balance equations (M), phase equilibrium relation (E), mole fraction summation for each stage (S) and energy balance equations (H), better known as MESH equations.

In this work, the traditional Isothermal Sum Rate (ISR) method (Tsuboka & Katayama, 1976) has been considered for the stage-wise calculation. We have tried to obtain the optimum number of stages and solvent flow rate by minimizing the multi-stage extractor cost for extraction of butanol and ethanol from aqueous solution using ILs: [TDTHP][DCA] and [TDTHP][Phosph].

6.2 Computational Details

6.2.1 Isothermal Sum Rate Algorithm

A hypothetical column (Figure 6.1) is considered in the model to formulate the optimization problem. The solvent enters at the bottom and feed enters at the top of the tray. The raffinate and extract compositions leaving each tray are assumed to be in equilibrium. Figure 6.2 shows the algorithm for the Tsuboka–Katayama ISR method (Tsuboka & Katayama, 1976). The extractor model is considered isothermal as stream temperature is uniform and heat of mixing is negligible. The number of stages (N) and solvent (IL) flow rate (Sol) are input variables for the ISR algorithm that are decided by the PSO algorithm described in the next section. The feed (water and solutes) flow rate is taken as 100 kmol/hr. The feed compositions are 0.99767 mole fraction water, 0.002 mole fraction butanol and 0.00033 mole fraction ethanol, which are identical with output composition of the acetone–butanol–ethanol (ABE) fermentation process. Initially the feed, extract and raffinate flow rates and

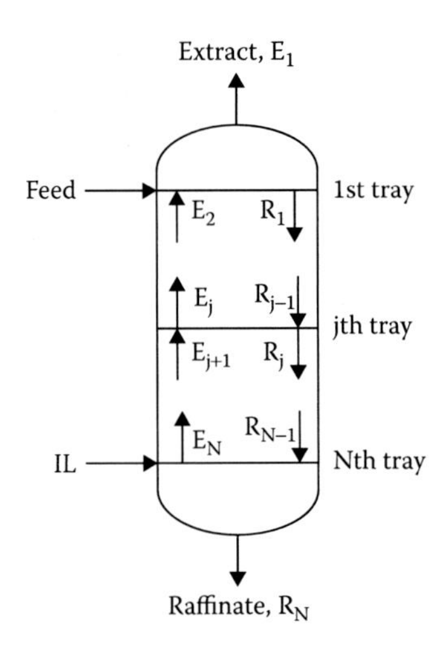

FIGURE 6.1
Hypothetical multi-stage liquid–liquid extractor.

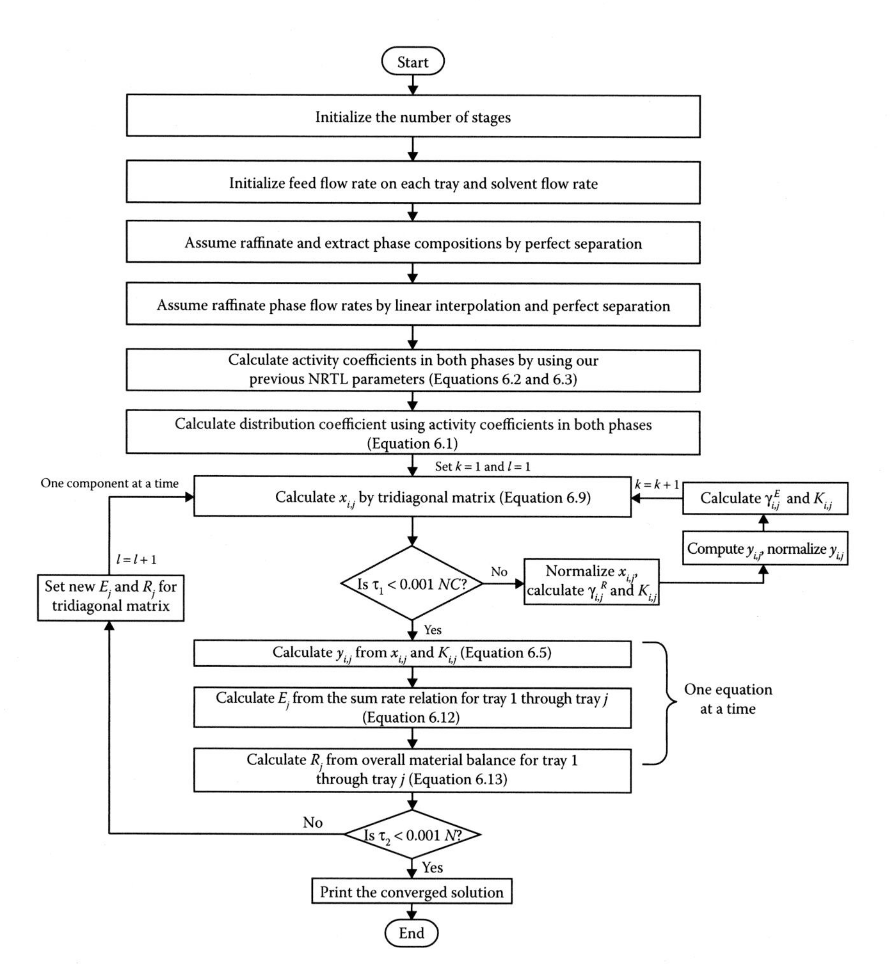

FIGURE 6.2
Tsuboka–Katayama ISR algorithm for liquid–liquid extraction.

compositions on each tray are derived based on linear interpolation with the assumption of complete separation.

The solvent is also assumed to be completely immiscible with water. The activity coefficients for each component (i) in both phases over every stage are then obtained from the nonrandom two liquid (NRTL) model. Binary interaction parameters (BIP) are derived from the NRTL model and are directly regressed from the experimental work. BIP for both quaternary systems are shown in Table 2.2 (Chapter 2). The distribution coefficients are then obtained by Equation 6.1.

$$K_{i,j} = \frac{\gamma_{i,j}^R}{\gamma_{i,j}^E} \tag{6.1}$$

Here the superscripts E and R indicate the extract (solvent-rich) and raffinate (water-rich) phases, respectively and $\gamma_{i,j}$ is the activity coefficient of component i in the respective phase over the jth tray. Here $\gamma_{i,j}$'s are obtained from the BIP values, that is,

$$\gamma_{i,j}^{R} = f\left(\tau_{ia}, x_{i,j}, T\right) \tag{6.2}$$

and

$$\gamma_{i,j}^{E} = f\left(\tau_{ia}, y_{i,j}, T\right) \tag{6.3}$$

where $T = 298.15$ K. $x_{i,j}$ and $y_{i,j}$ represent the compositions of component i in the raffinate and extract phases, respectively, from stage j. τ_{ia} is the BIP for a pair of components i and a. The overall material balance for component i over the jth tray is then given by

$$R_{j-1}x_{i,j-1} + E_{j+1}y_{i,j+1} + F_j z_{i,j} - R_j x_{i,j} - E_j y_{i,j} = 0 \tag{6.4}$$

where subscripts i and j represent the component and the tray, respectively. E and R represent the extract and raffinate flow rate, respectively, in kmol/hr from a specific tray. F stands for the feed flow rate in kmol/hr over a specific tray. The equilibrium relation for component i over j is given by

$$y_{i,j} = K_{i,j}x_{i,j} \tag{6.5}$$

From Equations 6.4 and 6.5, we have

$$R_{j-1}x_{i,j-1} - \left(R_j + E_j K_{i,j}\right)x_{i,j} + E_{j+1}K_{i,j+1}x_{i,j+1} = -F_j z_{i,j} \tag{6.6}$$

For the top ($j = 1$) and bottom ($j = N$) stages, the raffinate (R_0) and extract (E_{N+1}) flow rates are zeros, respectively. So the simplified equations can be written as

$$-\left(R_1 + E_1 K_{i,1}\right)x_{i,1} + E_2 K_{i,2}x_{i,2} = -F_1 z_{i,1} \tag{6.7}$$

$$R_{N-1}x_{i,N-1} - \left(R_N + E_N K_{i,N}\right)x_{i,N} = -F_N z_{i,N} \tag{6.8}$$

Equations 6.6 through 6.8 can be arranged in the form of a tridiagonal matrix,

$$\begin{pmatrix} B_1 & C_1 & 0 & 0 & 0 & 0 \\ A_2 & B_2 & C_2 & 0 & 0 & 0 \\ \cdots & \cdots & \cdots & \cdots & \cdots & \cdots \\ 0 & 0 & A_j & B_j & C_j & 0 \\ \cdots & \cdots & \cdots & \cdots & \cdots & \cdots \\ 0 & 0 & 0 & A_{N-1} & B_{N-1} & C_{N-1} \\ 0 & 0 & 0 & 0 & A_N & B_N \end{pmatrix} \begin{pmatrix} x_{i,1} \\ x_{i,2} \\ \cdots \\ x_{i,j} \\ \cdots \\ x_{i,N-1} \\ x_{i,N} \end{pmatrix} = \begin{pmatrix} D_1 \\ D_2 \\ \cdots \\ D_j \\ \cdots \\ D_{N-1} \\ D_N \end{pmatrix} \tag{6.9}$$

where $A_j = R_{j-1}$, $B_j = -(R_j + E_jK_{i,j})$, $C_j = E_{j+1}K_{i,j+1}$, $D_j = -F_jz_{i,j}$ for $j = 1$ to N tray and component i. The tridiagonal matrix for each component is solved by the direct method for a sparse linear system in MATLAB®.

The new values of $x_{i,j}$ are compared with assumed values as per the inner loop termination criterion (Equations 6.10 and 6.11).

$$\tau_1 = \sum_{j=1}^{N}\sum_{i=1}^{C}\left|x_{i,j}^{(r-1)} - x_{i,j}^{r}\right| \tag{6.10}$$

where N and C represent the total number of trays and total number of components, respectively. k is the inner loop index.

$$\tau_1 < 0.01NC \tag{6.11}$$

The new $x_{i,j}$ values are normalized and then used to compute the activity coefficient in raffinate phase ($R_{i,j}$) by Equation 6.2. Thereafter $K_{i,j}$ and $y_{i,j}$ are computed as per Equations 6.1 and 6.5, respectively. The new $y_{i,j}$ values are again normalized and then used to compute $E_{i,j}$ (Equation 6.3) and $K_{i,j}$ (Equation 6.1). The tridiagonal matrix is then solved with these new $K_{i,j}$ values till the termination criterion is satisfied (Equations 6.10 and 6.11). After the convergence of inner loop, $x_{i,j}$ values are utilized to calculate $y_{i,j}$ (Equation 6.5) using the new values of activity coefficients. New E_j values are then calculated from the sum rate relation

$$E_j^{(l+1)} = E_j^{l}\sum_{i=1}^{C} y_{i,j} \tag{6.12}$$

where l is the outer loop index. The outer loop termination criterion is considered as

$$\tau_2 = \sum_{j=1}^{N}\left(1 - \frac{E_j^{l}}{E_j^{(l+1)}}\right)^2 < 0.001N \tag{6.13}$$

For the diverging solution, the corresponding R_j values are obtained as

$$R_j = E_{j+1} - E_1 + \sum_{1}^{j} F_j \tag{6.14}$$

The tridiagonal matrix with new extract and raffinate flow rates is now solved till the outer loop gets converged. The converged solution then consists of tray-wise composition and flow rates of extract and raffinate leaving a particular tray. The solute and solvent amounts in the final raffinate stream are then used as variables in the cost (objective) function.

6.2.2 Particle Swarm Optimization Algorithm

The PSO is a stochastic optimization method based on swarm intelligence. It was first proposed by Kennedy (Eberhart & Kennedy, 1995a, 1995b) recognizing the flocking behavior of birds as the principle for optimization. Each particle in population (swarm) is represented by two vectors, namely position and velocity. These vectors are updated by the past experience of particles and their neighbors. Inertial, cognitive and social components that play a major role in effectiveness and performance of PSO update the vectors iteratively. In the literature, there are many modified versions of PSO applied to various domains. The PSO algorithm is explained in Figure 6.3.

Each variable i is represented by *nop* (number of populations)-dimensional position and velocity vectors. Both vectors are initialized randomly within the search space. The corresponding objective function to each population is then calculated. The best position of variable i through the generation cycle is known as the individual best position ($p_{best,i}$), while the position of the best variable in its entire population is termed as the global best position (g_{best}). The best positions are usually decided based on the minimum objective function values. The position (X_i) and velocity (V_i) of particle i at iteration $k + 1$ are updated by

$$V_i^{k+1} = \omega^k V_i^k + c_1 r_1 \left(p_{\text{best},i}^k - X_i^k\right) + c_2 r_2 \left(g_{\text{best}}^k - X_i^k\right) \tag{6.15}$$

$$X_i^{k+1} = X_i^k + V_i^{k+1} \tag{6.16}$$

where c_1 and c_2 are learning factors representing the stochastic acceleration term weighting. Generally, $c_1 = c_2 = 2$ and r_1 and r_2 are random numbers generated separately from 0 to 1. $p_{\text{best},i}^k$ represents the best position of variable i till the kth iteration, while g_{best}^k is the best global position in swarm till the kth iteration. ω is the inertia weight term providing balance between global and local exploration ability. Among the various inertia term mechanisms proposed by different authors, a simple mechanism with a Linear Decreasing Inertia Weight (LDIW) is proposed as follows:

$$\omega^k = \omega_{\max} - \frac{\omega_{\max} - \omega_{\min}}{\text{iter}_{\max}} \times k \tag{6.17}$$

where $\omega_{\max} = 0.9$, $\omega_{\min} = 0.4$. $\text{iter}_{\max}$ is the maximum number of iterations decided by the user.

Arumugam and Rao (2008) proposed another strategy which is the inertia weight and acceleration coefficient based on global and local best values. The Global-average Local best Inertia Weight (GLbestIW) for variable i is

$$\omega_i = \left(1.1 - \frac{g_{\text{best}}}{\left(p_{\text{best},i}\right)_{\text{average}}}\right) \tag{6.18}$$

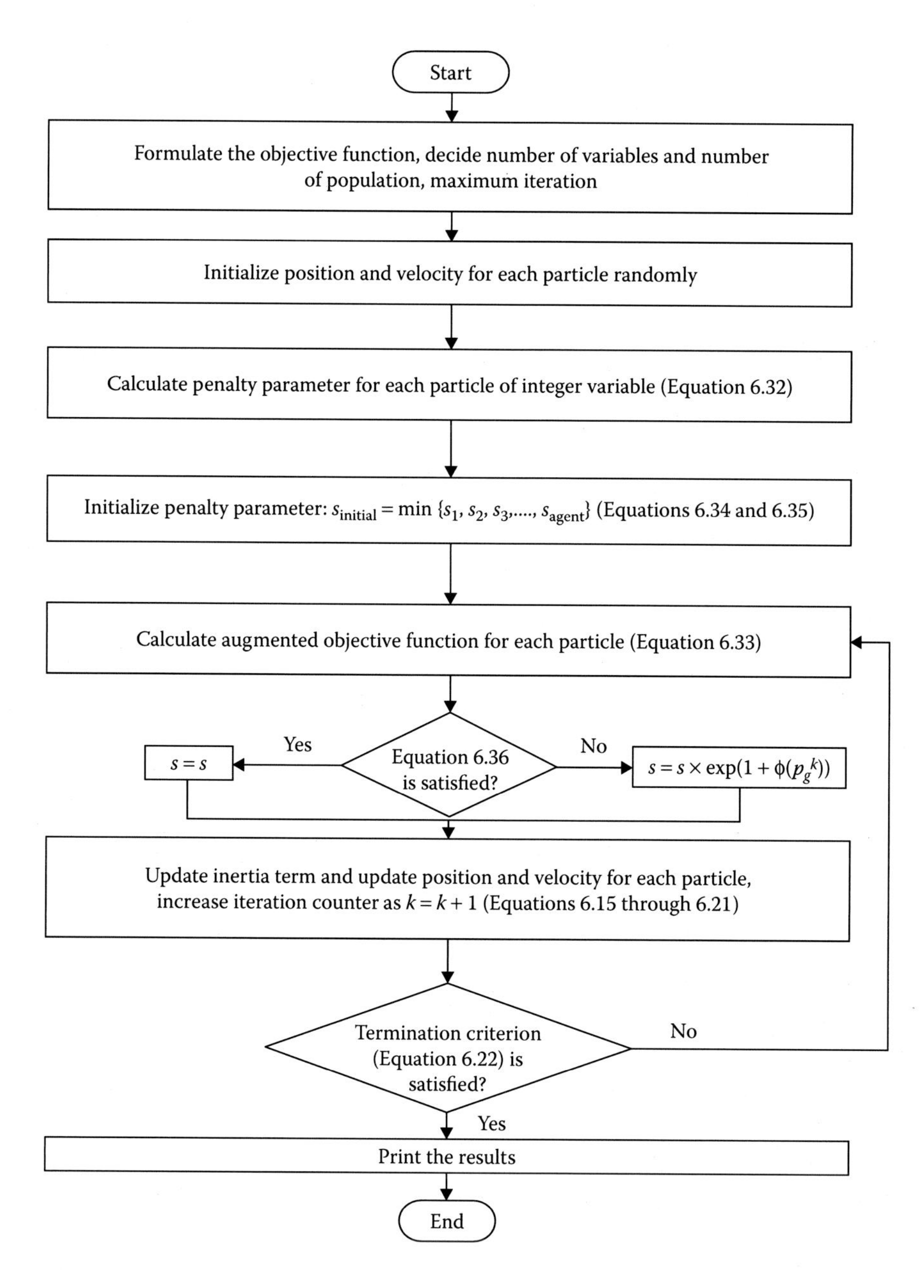

FIGURE 6.3
Particle swarm optimization algorithm.

and the Global-Local best Acceleration Coefficient (GLbestAC) is

$$\mathrm{GLbestAC} = \left(1 + \frac{g_{\mathrm{best}}}{p_{\mathrm{best},i}}\right) \tag{6.19}$$

Here the ($p_{\mathrm{best},i}$) average is the average of all the personal best values in specific generation. The velocity (V_i) of particle i is updated by

$$V_i^{k+1} = \omega_i^k V_i^k + \mathrm{GLbestAC} \times r\left(p_{\mathrm{best},i}^k + g_{\mathrm{best}}^k - 2X_i^k\right) \tag{6.20}$$

where r is the random number generated from 0 to 1. The updated velocity from Equation 6.20 is then used to calculate the new position of the particle.

The position and velocity bounds are applied to the updated vectors to keep the particles within the search space. Position bounds are then decided from the problem variable bounds. The velocity is therefore clamped within $[-V_{\mathrm{max}}, V_{\mathrm{max}}]$, where V_{max} is given by

$$V_{\mathrm{max}} = \left(X_{UB} - X_{LB}\right)/2 \tag{6.21}$$

where X_{UB} and X_{LB} represent the upper and lower limit for variable X, respectively. Velocity bounds can be varied based on the necessity of the problem. Objective functions are then evaluated at updated positions and compared with past function values.

The best values ($p_{\mathrm{best},i}^k$ and g_{best}^k) are improved upon continuously with each iteration. The optimization continues till the termination criterion is met. The termination criterion can be of the maximum number of iterations. Generally, the termination criterion is given by

$$\sum_{n=1}^{nop} \frac{\left(X_n^{k+1} - X_n^k\right)^2}{nop} \le 10^{-4} \tag{6.22}$$

where n is the number of variables and nop is the number of populations.

6.2.3 Problem Formulation

The current work emphasizes on the cost optimization for the multi-stage liquid–liquid extractor with the implementation of the PSO and ISR algorithms. The annualized cost (C_{Total}) is considered as an objective function that contains three parameters, namely (1) solvent lost in final raffinate (C_{Solvent}), (2) solute/solutes lost in final raffinate (C_{Solute}) and (3) column capital cost (C_{Capital}).

$$C_{\mathrm{Total}} = C_{\mathrm{Solvent}} + C_{\mathrm{Solute}} + C_{\mathrm{Capital}} \tag{6.23}$$

The costs for solvent and solute loss are defined as

$$C_{\text{Solvent}} = R_N x_{BN}(\text{kmol/hr}) \times (24 \times 325)(\text{hrs/year}) \times MW_B \times P_B(\text{INR/kg}) \quad (6.24)$$

$$C_{\text{Solute}} = R_N x_{CN}(\text{kmol/hr}) \times (24 \times 325)(\text{hrs/year}) \times MW_C \times P_C(\text{INR/kg}) \quad (6.25)$$

where R_N, x_{BN} and x_{CN} are the raffinate flow rate and the solvent (*B*) and solute (*C*) composition in the raffinate phase from stage *N*, respectively. *MW* stands for molecular weight. *P* stands for the price of the component. Here ethanol and 1-butanol have been considered as solutes. The capital cost contains two components: packing cost (C_{pack}) and column cost (C_{col}). They are given as follows:

$$C_{\text{pack}} = \left(\frac{\pi}{4} D^2\right) \times N \times \text{HETS} \times C^0_{\text{pack}} \quad (6.26)$$

where *D*, *N*, HETS and C^0_{pack} are column diameter (m), total number of stages, height equivalent to theoretical stage (m) and packing cost per unit volume (INR/m^3), respectively. The column cost is given as follows:

$$C_{\text{col}} = 1.4 \times \pi \times D \times N \times \text{HETS} \times T_S \times \rho_S \times C_S \quad (6.27)$$

where T_S, ρ_S and C_S are the column thickness (m), steel (column material) density (kg/m^3) and steel cost per unit mass (INR/kg), respectively. Table 6.1 shows the values of different parameters in cost equations. With consideration

TABLE 6.1

Cost Function Parameters Values

Parameters	Value	Unit
Feed flow rate (*F*)	100	kmol/hr
Column diameter (*D*)	1*	m
HETS	1*	m
Packing cost (C^0_{pack})	11636.43*	INR/m^3
Column thickness (T_s)	0.006*	m
Steel density (ρ_s)	8000*	kg/m^3
Steel cost (C_s)	296.39*	INR/kg
[TDTHP][Phosph] cost	136172.40**	INR/kg
[TDTHP][DCA] cost	237480**	INR/kg
Butanol cost	744.76**	INR/kg
Ethanol cost	2210.39**	INR/kg

*Taken from Ubaidullah, S. et al., *J. Inst. Eng. E*, 93, 49–54, 2013.

**Taken from the Sigma Aldrich Online price catalogue as on August 7, 2014.

of additional costs due to depreciation, interest and maintenance, 35% excess amount in terms of capital cost has been set to the final capital cost.

$$C_{\text{capital}} = 1.35 \times \left(C_{\text{pack}} + C_{\text{col}}\right) \tag{6.28}$$

It was found that the objective function is highly nonlinear in nature. The number of stages and solvent flow rate have been considered as decision variables.

The optimization problem is then defined as

$$Obj_C = C_{\text{Total}} \tag{6.29}$$

with bounds

$$2 < N < 10 \tag{6.30}$$

$$1 < Sol < 40 \tag{6.31}$$

where N and Sol are the number of stages and solvent flow rate in kmol/hr, respectively. The bounds limit has been decided based on our known experimental data of Chapter 2. Here, variable N is an integer variable, while Sol is a continuous variable. So, the overall problem formulation is a Mixed Integer Nonlinear Programming (MINLP). Since one variable (N) is integer, the penalty function for the integer variable is used in the optimization problem. For the discrete variable, penalty functions in the form of sine and elliptic are widely used. Shin, Gürdal and Griffin (1990) showed that *elliptic*-type functions are unstable as compared to *sine*-type penalty functions. So, in this chapter, we have implemented the *sine*-type function (Shin et al., 1990) for the single integer variable N.

$$\phi(x) = \frac{1}{2}\left[\sin \frac{2\pi\left\{x_m^c - \frac{1}{4}\left(d_{j+1} + 3d_j\right)\right\}}{d_{j+1} - d_j} + 1\right] \tag{6.32}$$

where x_m^c is the continuous design variable between discrete variables, d_j and d_{j+1}. Thus implementing the penalty approach in optimization, the augmented objective function is defined as

$$F(x) = Obj + s\phi(x) \tag{6.33}$$

where s is the penalty parameter. The initial values of s are considered as

$$s_{\text{initial}} = \min\{s_1, s_2, \cdots\cdots, s_{nop}\} \tag{6.34}$$

where *nop* is the total number of populations and s is defined as

$$s_i = 1 + \phi(x_i) \text{ where } i = 1, 2, \cdots\cdots, nop \tag{6.35}$$

The penalty parameter through generation is updated based on tolerance (ε), where ε is a small positive number (=1). The termination criterion then is defined as

$$\frac{\left|F\left(g_{\text{best}}^{k}\right) - Obj_{C}\left(g_{\text{best}}^{k}\right)\right|}{F\left(g_{\text{best}}^{k}\right)} \leq \varepsilon \tag{6.36}$$

If Equation 6.36 is satisfied, s would be the same as the old value. Otherwise it is updated as per Equation 6.37 (Kitayama, Arakawa, & Yamazaki, 2006).

$$s = s \times \exp\left(1 + \phi\left(g_{\text{best}}^{k}\right)\right) \tag{6.37}$$

The optimization problem can then be rewritten in the following form:

$$\min \rightarrow F(N, Sol) \tag{6.38}$$

$$2 < N < 10 \tag{6.30}$$

$$1 < Sol < 40 \tag{6.31}$$

The PSO algorithm with the penalty function is shown in Figure 6.3. The whole optimization strategy is coded in MATLAB (Figure 6.4). The decision (Unknown) variables N and *Sol* were particles in swarm (Population). The PSO algorithm then invokes the cost function (Equation 6.33). The cost function thereafter calculates the capital cost using particle values. For the solvent and solute loss cost, the ISR algorithm is used. The ISR algorithm then calculates the stage-wise composition and flow rates with the help of activity coefficients generated by the NRTL model (Equations 6.2 and 6.3).

For the NRTL model, we use our own BIP (Table 2.2) measured in our previous experimental work. The PSO algorithm then executes and minimizes the augmented cost function (Equation 6.33) till the minimum sum of square errors for all particles is achieved.

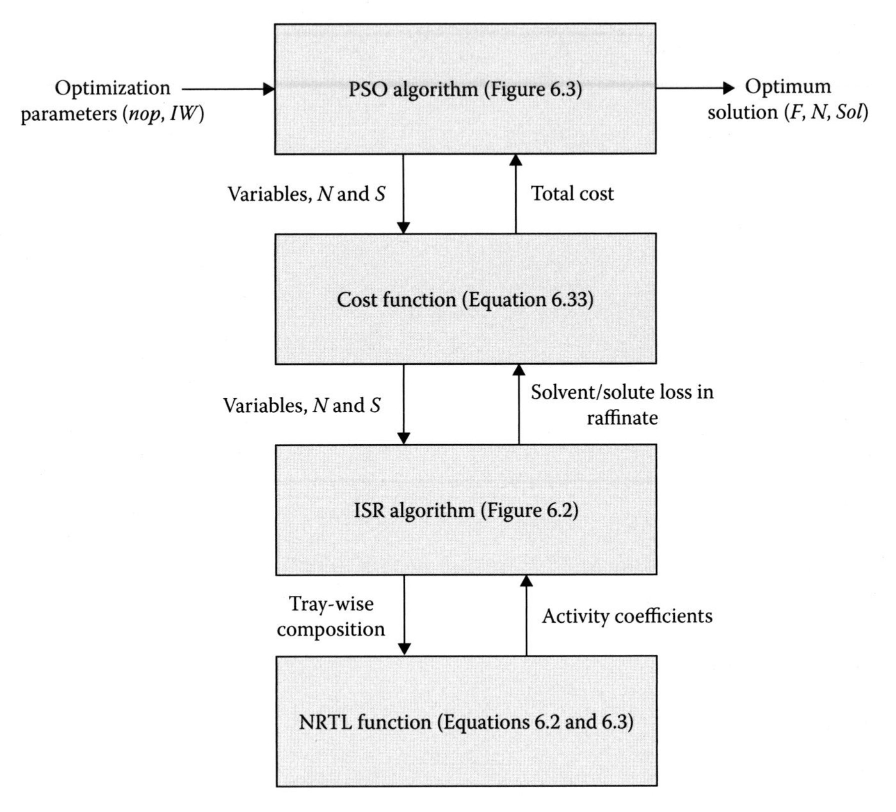

FIGURE 6.4
Optimization strategy with input/output variables.

6.3 Results and Discussions

6.3.1 Tuning of Particle Swarm Optimization Parameters

Since PSO is a stochastic algorithm, its efficiency is tuned by different parameters like population size and inertia weight (Equations 6.17 through 6.19, 6.32 and 6.33). Initially the effect of these parameters on PSO efficiency is checked (Figure 6.3). The efficiency is analysed by the evolution of different parameters like best solution, success rate (SR), average number of iterations (AIT), mean of the best solutions (mean) and standard deviation (SD). We consider 50 individual runs of optimization for the [TDTHP][DCA]–ethanol–1-butanol–water system in order to tune the parameters. The SR is defined as the percentage of runs giving the best solution. AIT is defined as the mean of total iterations to achieve the best solution. Mean and SD are calculated from the solutions across all 50 runs. The number of populations (*nop*) considered

TABLE 6.2

Parameter Tuning and Efficiency Analysis for the [TDTHP][DCA] (1)–Ethanol (2)–1-Butanol (3)–Water (4) System at $T = 298.15$ K and $p = 1$ atm

	GLbestIW		LDIW	
Population Size	**AIT**	**SR (%)**	**AIT**	**SR (%)**
10	44	34	100	70
20	41	48	100	82
30	42	64	99	86
40	39	64	100	90
50	41	82	100	94
60	40	74	100	92
70	43	70	100	94
80	37	78	100	96
90	40	80	100	98
100	40	76	100	98

for the study ranges from 10 to 100 with an increment of 10. Two inertia weight approaches (Equations 6.17 through 6.19) are studied for the optimization. Table 6.2 shows the SR and AIT for both inertia weight approaches with different populations for the [TDTHP][DCA]–ethanol–1-butanol–water system. LDIW shows the high success rate as compared to GLbestIW.

LDIW is found to decrease linearly independent of g_{best} and $p_{\text{best},i}$ values (Equation 6.17), thereby giving the solution after a higher number of generations. From Figures 6.5 through 6.9, it can be seen that few local minima are evaluated far away from global minima at a higher number of generations. It indicates that a higher number of generations are required to achieve the termination criterion (Equation 6.22). GLbestIW shows very few local minima at the initial generations (Figures 6.10 through 6.14). Thus both GLbestIW and GLbestAC depend on g_{best} and $p_{\text{best},i}$ values (Equations 6.18 and 6.19), which converge faster. AIT for LDIW was approximately 2.5 times higher than GLbestIW. Thus GLbestIW is chosen for further study considering lesser function evaluations. The population containing 10 and 20 particles shows SR less than 50% with GLbestIW, while the population of 30 and 40 particles has similar SR and AIT. Higher *nop* (>50) did not improve the SR and AIT significantly. So we choose a population of 30 particles for optimization as a fewer number of function evaluations are needed.

6.3.2 Cost Optimization Results

The objective function (Equation 6.33) is optimized for two systems, namely, [TDTHP][DCA]–ethanol–1-butanol–water and [TDTHP][Phosph]–ethanol–1-butanol–water. PSO converges at various best solutions with success rate as shown in Figure 6.15. The minimum objective function is found to be 1.1253×10^{11} INR/year with an SR of 64% for the [TDTHP][DCA]–ethanol–1-butanol–water

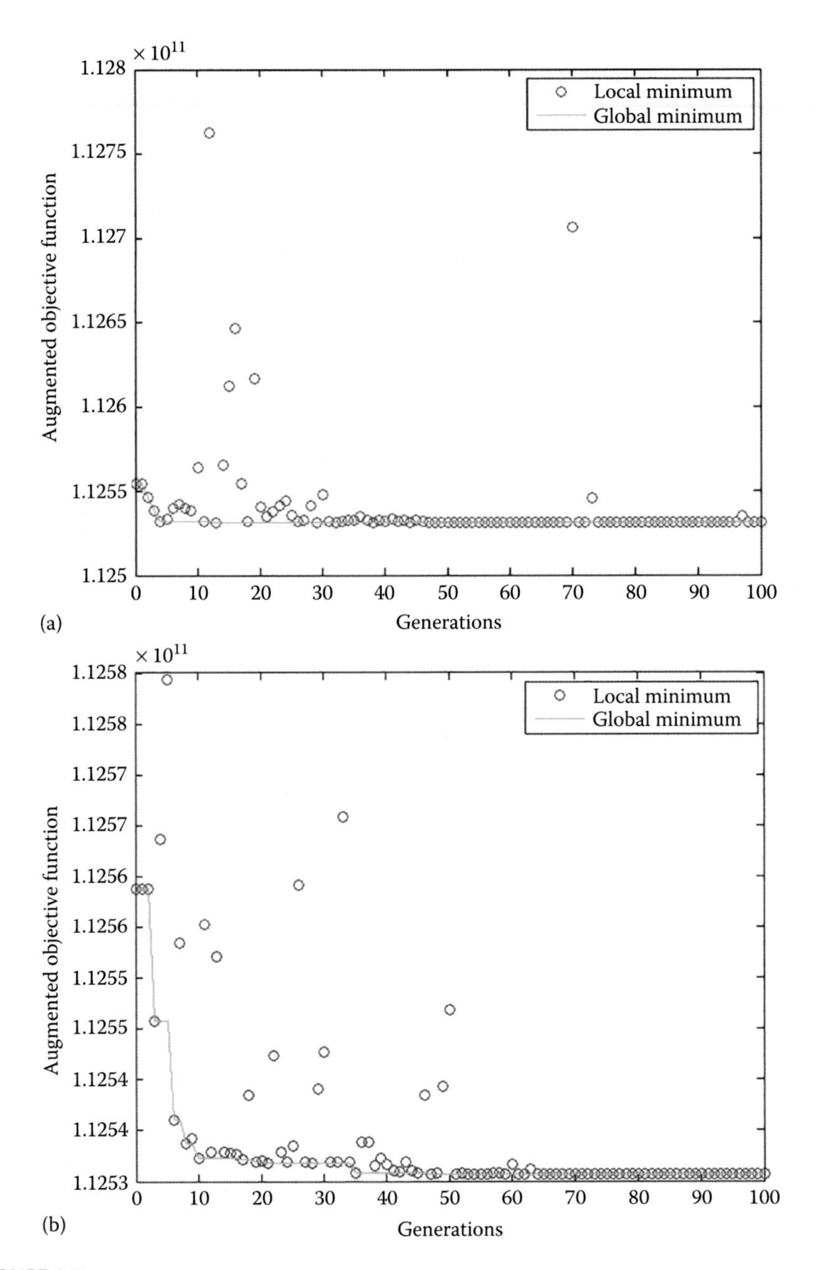

FIGURE 6.5
Augmented objective function (Equation 6.33) versus generations for LD inertia weight for the [TDTHP][DCA] (1)–ethanol (2)–1-butanol (3)–water (4) system at $T = 298.15$ K and $p = 1$ atm. (a) *nop* = 10 and (b) *nop* = 20.

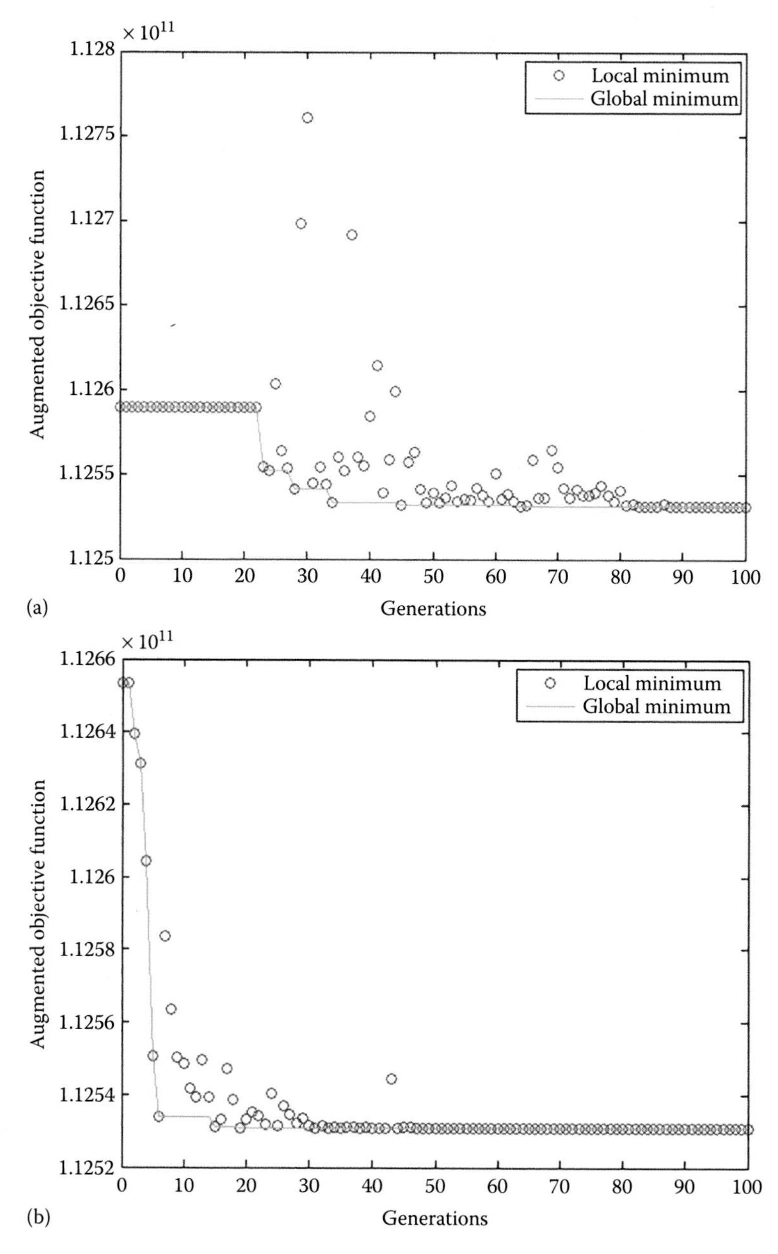

FIGURE 6.6
Augmented objective function (Equation 6.33) versus generations for LD inertia weight for the [TDTHP][DCA] (1)–ethanol (2)–1-butanol (3)–water (4) system at $T = 298.15$ K and $p = 1$ atm. (a) $nop = 30$ and (b) $nop = 40$.

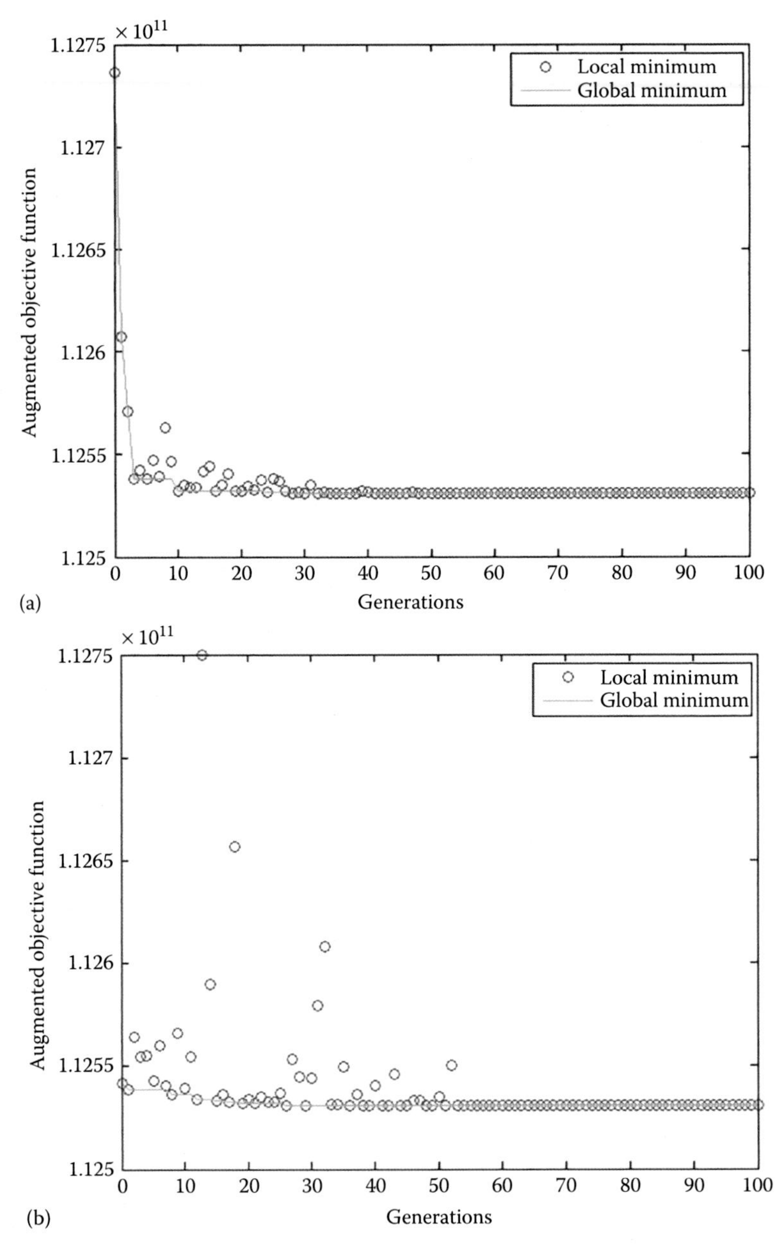

FIGURE 6.7
Augmented objective function (Equation 6.33) versus generations for LD inertia weight for the [TDTHP][DCA] (1)–ethanol (2)–1-butanol (3)–water (4) system at $T = 298.15$ K and $p = 1$ atm. (a) $nop = 50$ and (b) $nop = 60$.

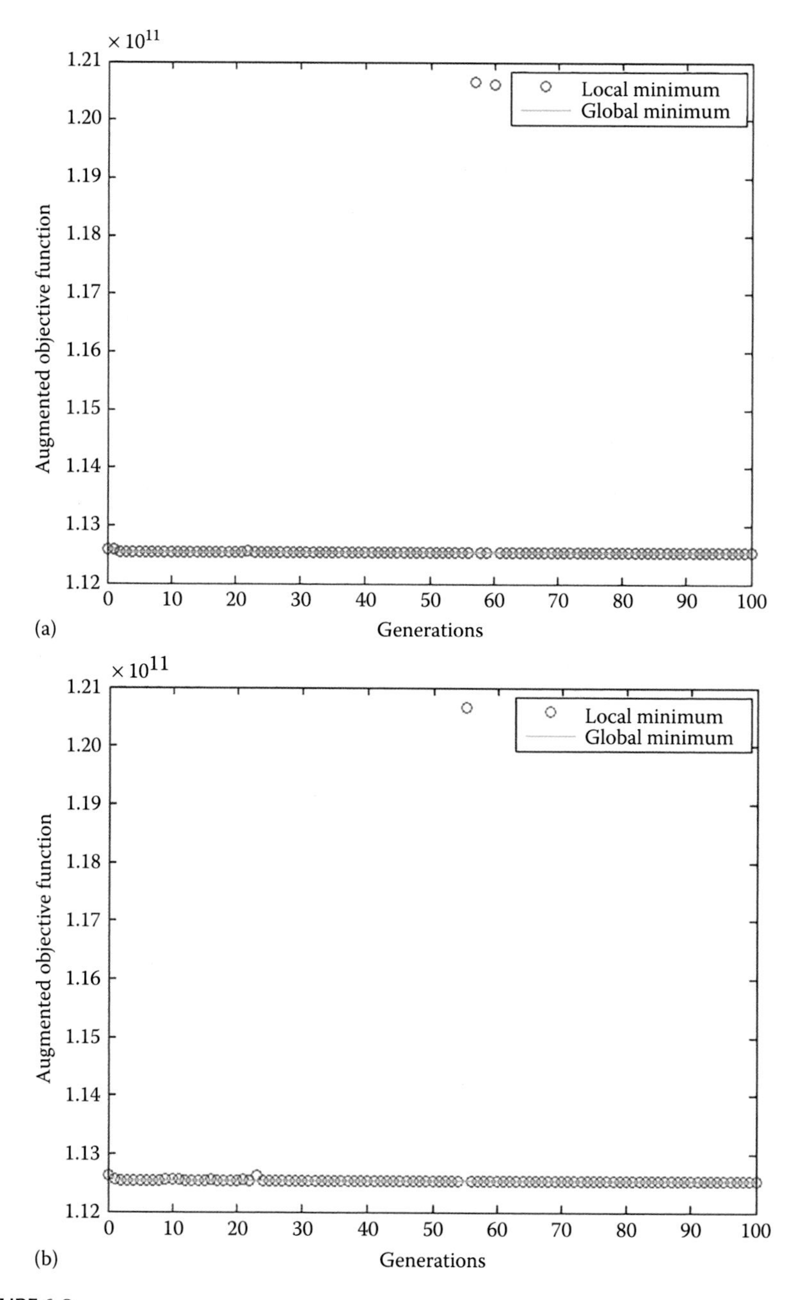

FIGURE 6.8
Augmented objective function (Equation 6.33) versus generations for LD inertia weight for the [TDTHP][DCA] (1)–ethanol (2)–1-butanol (3)–water (4) system at $T = 298.15$ K and $p = 1$ atm. (a) $nop = 70$ and (b) $nop = 80$.

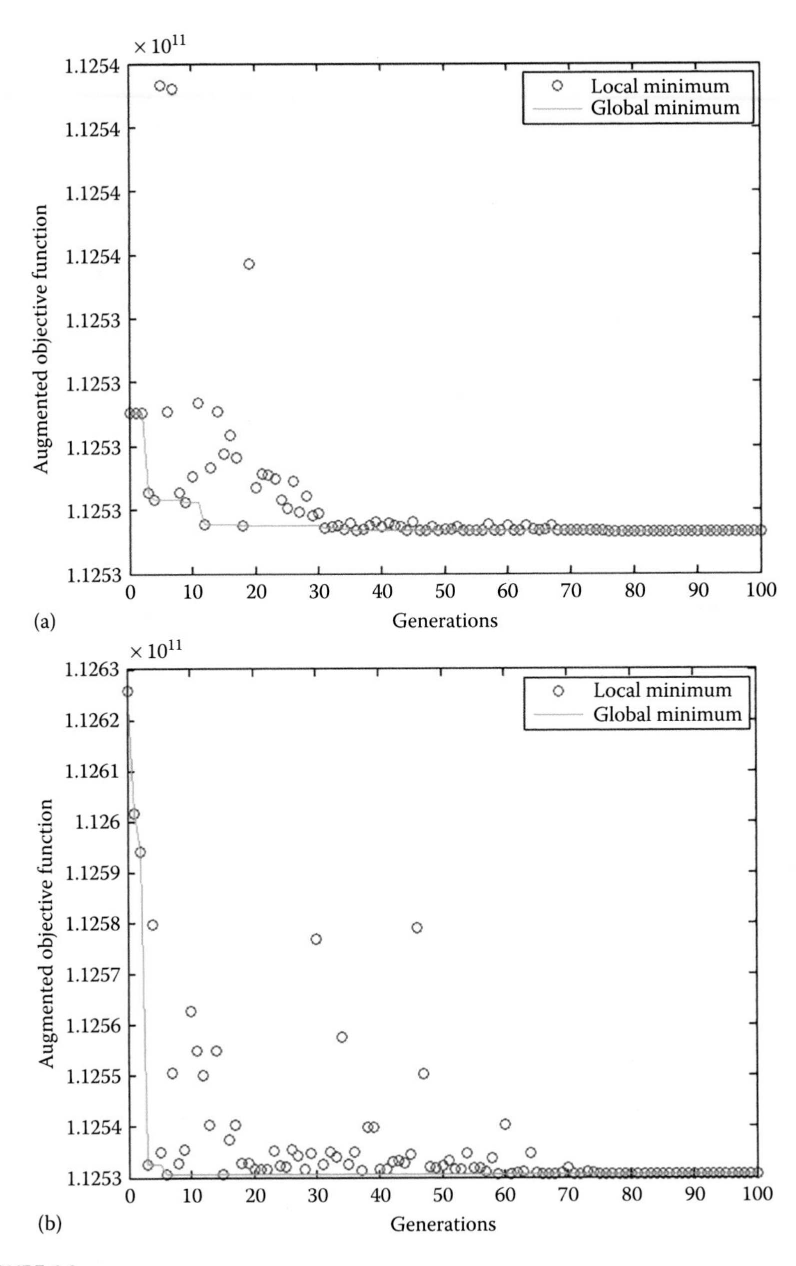

FIGURE 6.9
Augmented objective function (Equation 6.33) versus generations for LD inertia weight for the [TDTHP][DCA] (1)–ethanol (2)–1-butanol (3)–water (4) system at $T = 298.15$ K and $p = 1$ atm. (a) $nop = 90$ and (b) $nop = 100$.

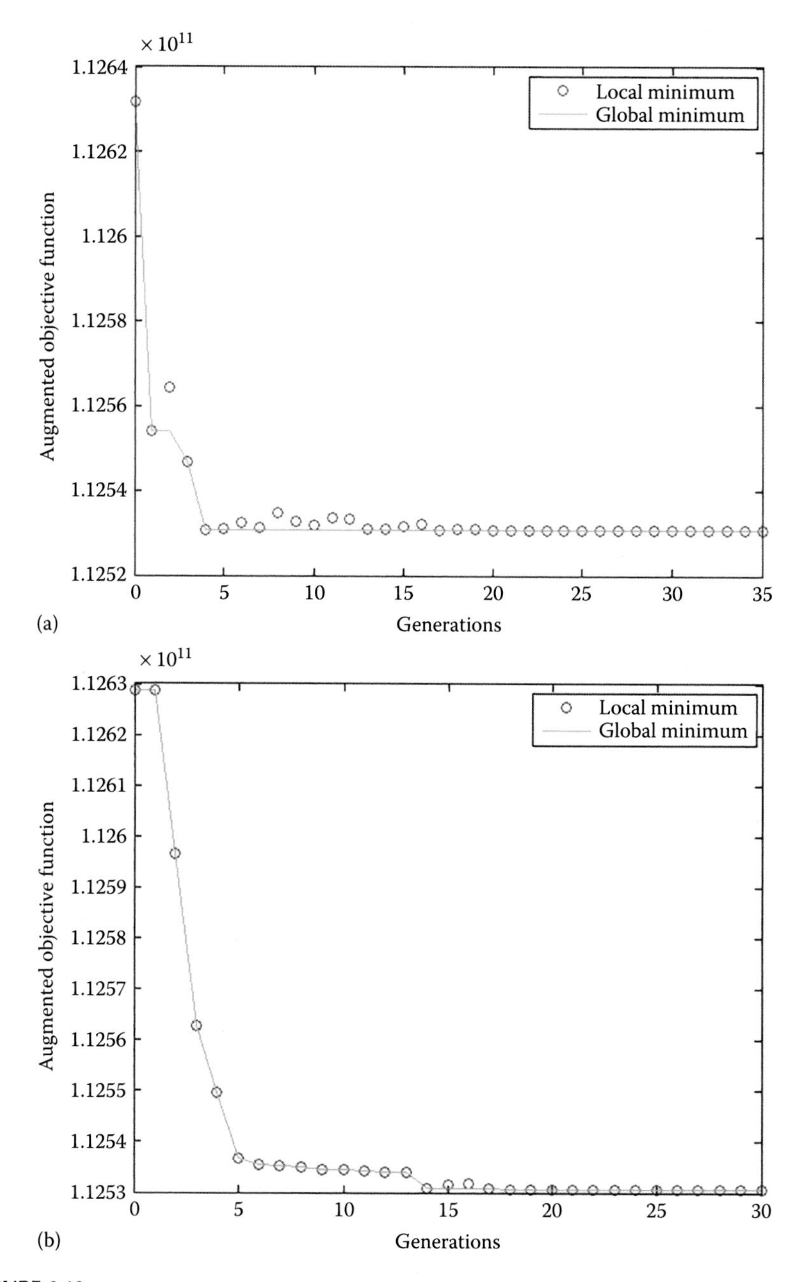

FIGURE 6.10
Augmented objective function (Equation 6.33) versus generations for GLBest inertia weight for the [TDTHP][DCA] (1)–ethanol (2)–1-butanol (3)–water (4) system at $T = 298.15$ K and $p = 1$ atm. (a) $nop = 10$ and (b) $nop = 20$.

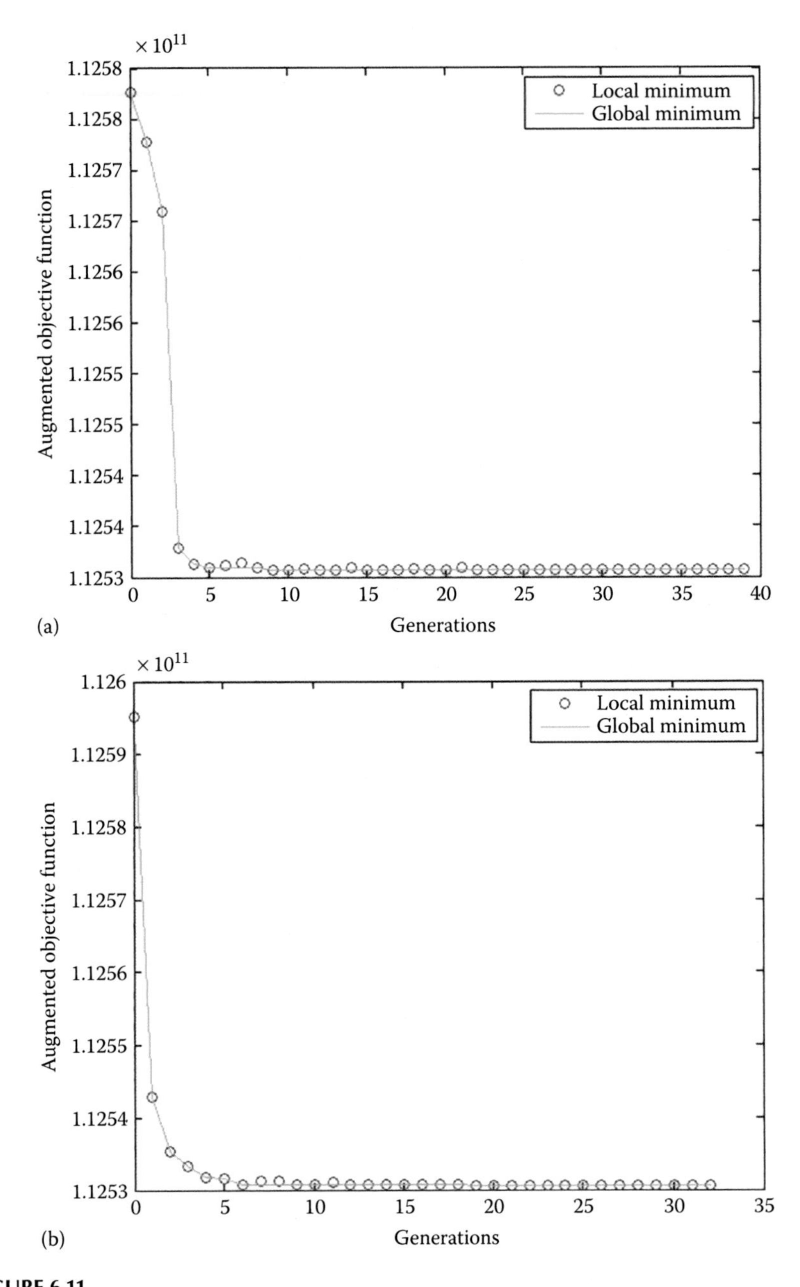

FIGURE 6.11
Augmented objective function (Equation 6.33) versus generations for GLBest inertia weight for the [TDTHP][DCA] (1)–ethanol (2)–1-butanol (3)–water (4) system at $T = 298.15$ K and $p = 1$ atm. (a) $nop = 30$ and (b) $nop = 40$.

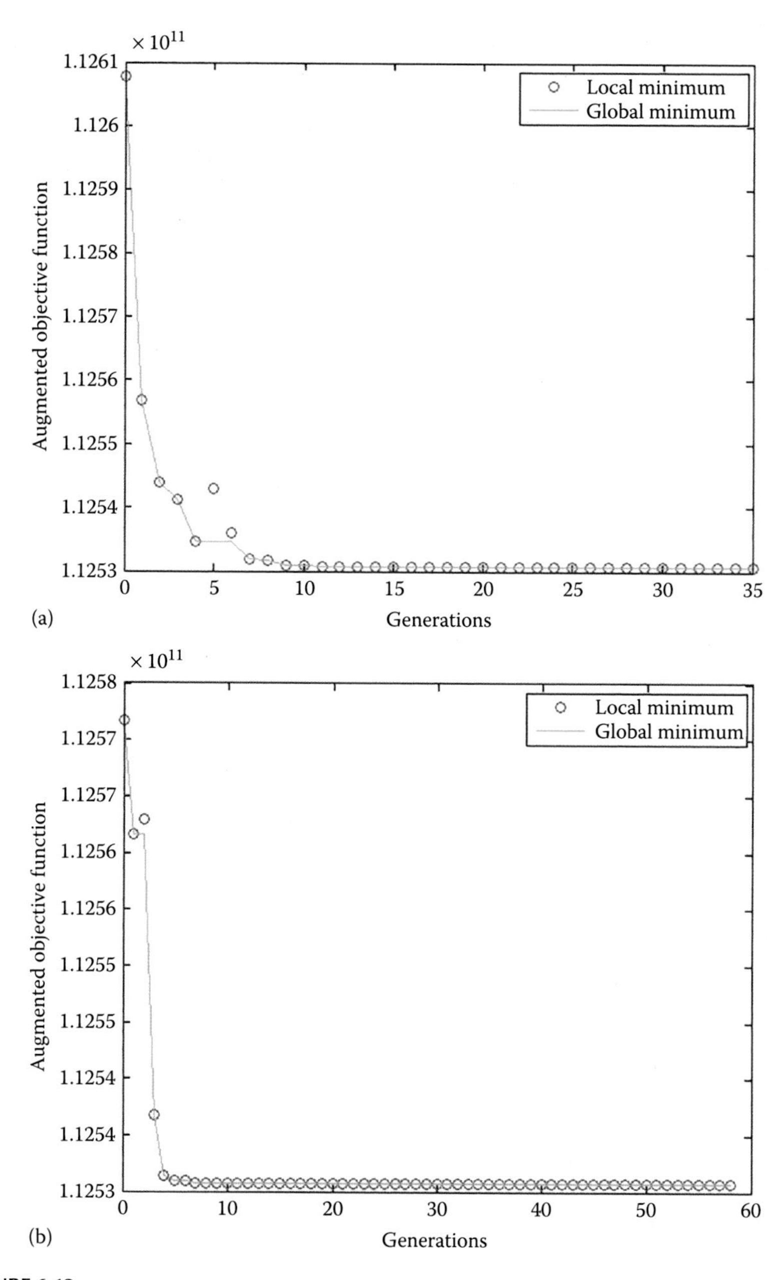

FIGURE 6.12
Augmented objective function (Equation 6.33) versus generations for GLBest inertia weight for the [TDTHP][DCA] (1)–ethanol (2)–1-butanol (3)–water (4) system at $T = 298.15$ K and $p = 1$ atm. (a) $nop = 50$ and (b) $nop = 60$.

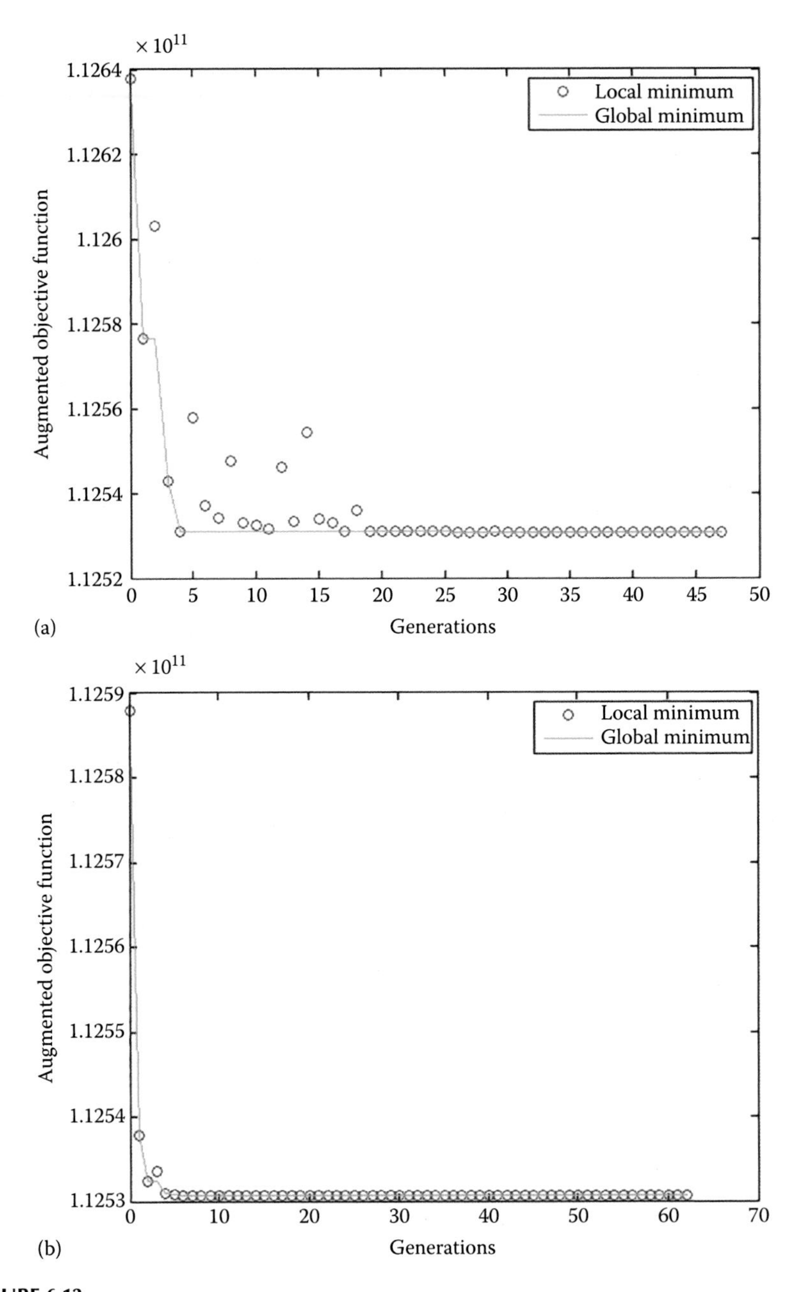

FIGURE 6.13
Augmented objective function (Equation 6.33) versus generations for GLBest inertia weight for the [TDTHP][DCA] (1)–ethanol (2)–1-butanol (3)–water (4) system at $T = 298.15$ K and $p = 1$ atm. (a) $nop = 70$ and (b) $nop = 80$.

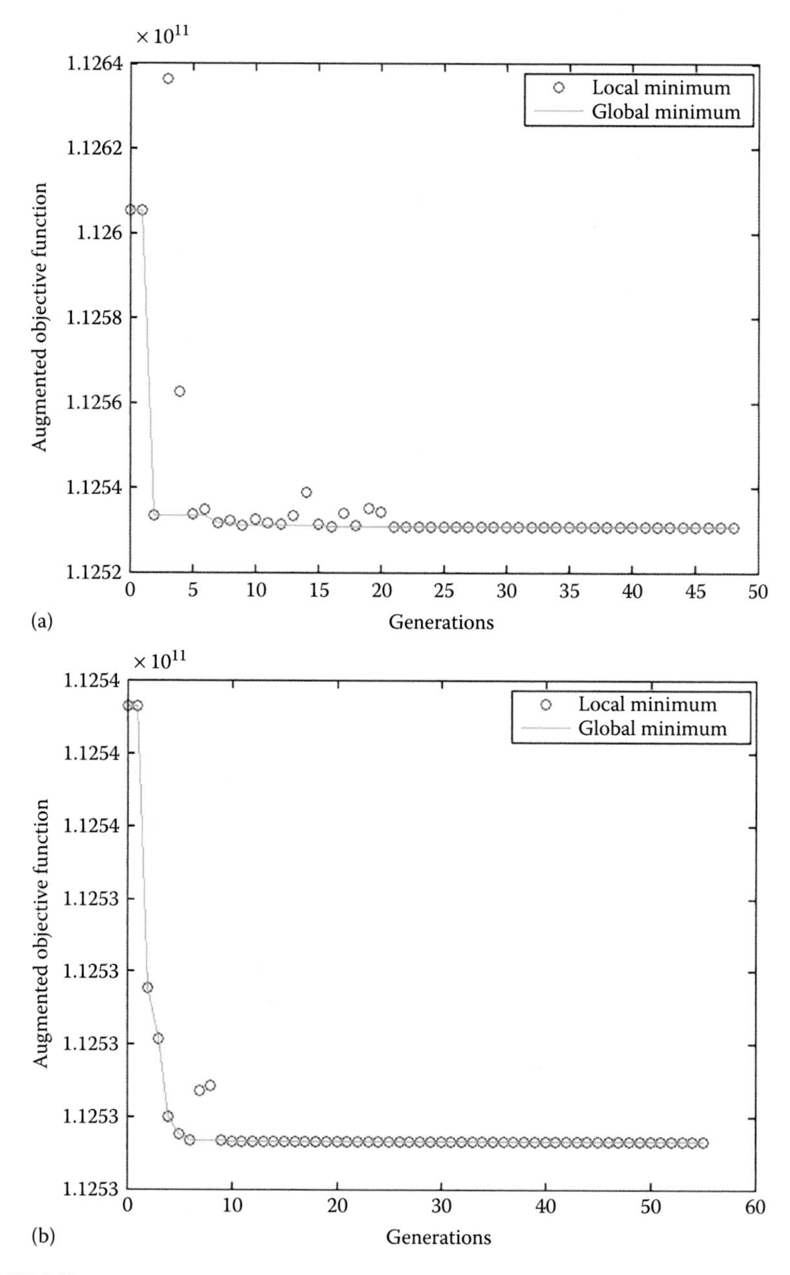

FIGURE 6.14
Augmented objective function (Equation 6.33) versus generations for GLBest inertia weight for the [TDTHP][DCA] (1)–ethanol (2)–1-butanol (3)–water (4) system at $T = 298.15$ K and $p = 1$ atm. (a) $nop = 90$ and (b) $nop = 100$.

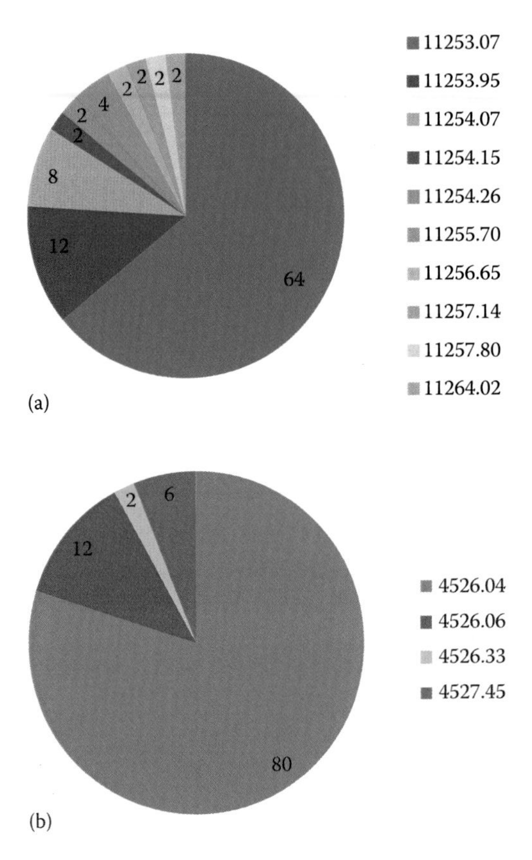

FIGURE 6.15
Optimized cost values (Equation 6.33) in crore INR/year with success rate (%) for the systems: (a) [TDTHP][DCA] (1)–ethanol (2)–1-butanol (3)–water (4) and (b) [TDTHP][Phosph] (1)–ethanol (2)–1-butanol (3)–water (4) at $T = 298.15$ K and $p = 1$ atm.

system. Nine local minimum solutions are also found with SR 36%. For the [TDTHP][Phosph]–ethanol–1-butanol–water system, the best solution is at 4.526×10^{10} INR/year with SR 80%. There are other three local solutions with a low SR of 20%. The optimized parameters corresponding to the best solution are shown in Table 6.3. For a system containing [TDTHP][DCA], the optimum number of stages and solvent flow rate are 3 and 3.9891 kmol/hr, respectively.

The optimum objective function value is found at the 23rd generation and the convergence criterion is fully satisfied at the 48th generation (Figure 6.16). The AIT over the 50 runs to achieve this solution is 42. Similarly for a [TDTHP][Phosph] containing system, the optimum solution is found with 10 stages and 36.285 kmol/hr solvent flow rate. It is achieved in 15 generations with AIT equal to 13 (Table 6.3 and Figure 6.17). The stage-wise compositions and flow rates with respect to the optimum solution for both systems are shown in Table 6.4. The [TDTHP][Phosph] containing system converges at an upper bound of N (=10).

TABLE 6.3

Optimization Results for the Multi-Stage Extractor at $T = 298.15$ K and $p = 1$ atm

Total Cost (INR/year)	N	*Sol* (kmol/hr)	Losses in Raffinate: IL Loss (kmol/hr)	Losses in Raffinate: Butanol Loss (kmol/hr)	Losses in Raffinate: Ethanol Loss (kmol/hr)	Optimization Evaluation Parameters: AIT	Optimization Evaluation Parameters: SR (%)	Optimization Evaluation Parameters: Mean (INR/year)	Optimization Evaluation Parameters: SD (INR/year)
[TDTHP][DCA] (1)–Ethanol (2)–1-Butanol (3)–Water (4)									
1.1253E+11	3	3.9891	0.1102	0.1183	0.0001	42	64	1.1254E+11	1.8330E+07
[TDTHP][Phosph] (1)–Ethanol (2)–1-Butanol (3)–Water (4)									
4.5260E+10	10	36.2850	0.0551	0.0375	0.0321	13	82	4.5261E+10	3.3884E+06

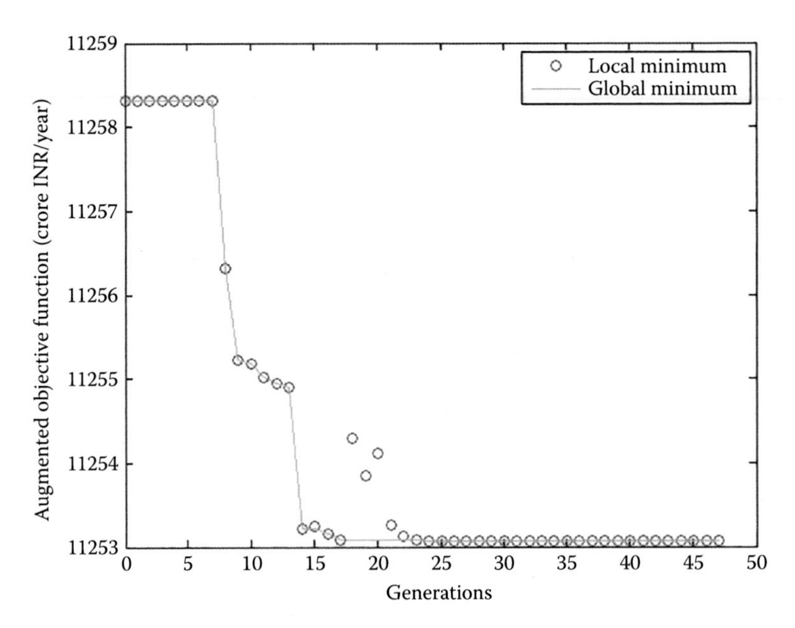

FIGURE 6.16
Optimized objective function (Equation 6.33) with iterations for the [TDTHP][DCA] (1)–ethanol (2)–1-butanol (3)–water (4) system at $T = 298.15$ K and $p = 1$ atm.

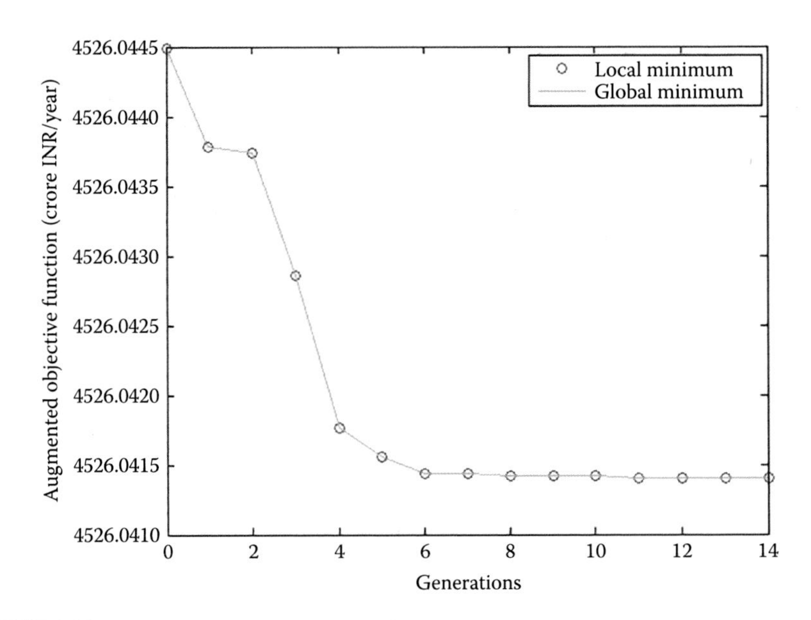

FIGURE 6.17
Optimized objective function (Equation 6.33) with iterations for the [TDTHP][Phosph] (1)–ethanol (2)–1-butanol (3)–water (4) system at $T = 298.15$ K and $p = 1$ atm.

TABLE 6.4

Stagewise Compositions and Flow Rates at $T = 298.15$ K and $p = 1$ atm

Stage (j)	Compositions*				Flow Rate (kmol/hr)
	x_W/y_W	x_B/y_B	x_E/y_E	x_{IL}/y_{IL}	
[TDTHP][Phosph] (1)–Ethanol (2)–1-Butanol (3)–Water (4)					
Extract Phase					
1	0.0020	0.0204	0.0082	0.9694	4.0014
2	0.0020	0.0179	0.0010	0.9791	4.0750
3	0.0020	0.0125	0.0001	0.9854	4.0489
Raffinate Phase					
1	0.9970	0.0019	0.0000	0.0011	100.0737
2	0.9972	0.0017	0.0000	0.0011	100.0476
3	0.9977	0.0012	0.0000	0.0011	99.9878
[TDTHP][DCA] (1)–Ethanol (2)–1-Butanol (3)–Water (4)					
Extract Phase					
1	0.0002	0.0045	0.0000	0.9953	36.4008
2	0.0002	0.0023	0.0000	0.9975	36.3792
3	0.0002	0.0012	0.0000	0.9986	36.3400
4	0.0003	0.0006	0.0000	0.9991	36.3195
5	0.0003	0.0003	0.0000	0.9994	36.3087
6	0.0003	0.0001	0.0000	0.9996	36.3031
7	0.0003	0.0000	0.0000	0.9996	36.3006
8	0.0003	0.0000	0.0000	0.9996	36.2992
9	0.0003	0.0000	0.0000	0.9997	36.2974
10	0.0002	0.0000	0.0000	0.9998	36.2936
Raffinate Phase					
1	0.9979	0.0012	0.0003	0.0006	99.9785
2	0.9983	0.0008	0.0003	0.0006	99.9392
3	0.9985	0.0006	0.0003	0.0006	99.9188
4	0.9986	0.0005	0.0003	0.0006	99.9079
5	0.9987	0.0004	0.0003	0.0006	99.9024
6	0.9987	0.0004	0.0003	0.0006	99.8998
7	0.9987	0.0004	0.0003	0.0006	99.8985
8	0.9987	0.0004	0.0003	0.0006	99.8966
9	0.9987	0.0004	0.0003	0.0006	99.8929
10	0.9987	0.0004	0.0003	0.0006	99.8842

*x and y refer to the raffinate (water-rich) and extract (IL-rich) phase composition, respectively.

The revised upper bound (N = 20–40) shows no improvement to get the solution within the bound regions. Therefore considering four decimal places as significant and physics of the problem, the present solution is considered as the optimum solution. The butanol concentration is reduced from 0.002 mole fraction in the feed to 0.0012 mole fraction and 0.0004 mole fraction in the raffinate phase for the [TDTHP][DCA] and [TDTHP][Phosph] systems, respectively (Table 6.4).

The extraction efficiency for solute is given by

$$\text{Efficiency}(\%) = \frac{\text{Solute in extract, kmol/hr}}{\text{Solute in feed, kmol/hr}} \times 100 \tag{6.39}$$

[TDTHP][Phosph] gives butanol efficiency at 81.25%, which is double that achieved by [TDTHP][DCA] (40.85%). But [TDTHP][DCA] proves to be a better solvent for ethanol extraction (99.70% efficiency) than [TDTHP][Phosph] (2.7% efficiency). It indicates that [TDTHP][Phosph] is more selective to butanol than ethanol. In summary, [TDTHP][DCA] is a better solvent for simultaneous extraction of butanol and ethanol from aqueous solution. [TDTHP][Phosph] loss (0.0551 kmol/hr) in the raffinate phase is half that of [TDTHP][DCA] loss (0.1102 kmol/hr), as [TDTHP][Phosph] is more hydrophobic than [TDTHP][DCA] in the water-rich phase. It results in lesser total cost in extraction by [TDTHP][Phosph] even though possessing a higher number of stages.

6.3.3 Effect of Ionic Liquid Cost on Optimization

An attempt has been made to study the impact of the IL cost on cost function values as compared to other components. This is in keeping in mind that newer production technologies will reduce the IL cost in the future. Fifty optimization runs are performed with the reduction in IL cost ranging through factors 1/1 to 1/10000 of the present cost. For both systems, the optimized results are shown in Table 6.5 and Figures 6.18 and 6.19. It is observed that the total cost gradually decreases with IL cost reduction. [TDTHP][DCA] gives 100% butanol efficiency when the IL cost is reduced by 100 times.

The corresponding number of stages and solvent flow rate are 9 and 20.3997 kmol/hr, respectively. Further reduction in the IL cost (by 1000 and 10000 times) reduces the total cost with the same butanol efficiency. The optimized number of stages is then 5 and solvent flow rate is 40 kmol/hr. Similarly for [TDTHP][Phosph], 100 times reduction in the IL cost shows the butanol efficiency of 98%. It also indicates a further improvement in ethanol efficiency from 2.7% to 24%. For [TDTHP][Phosph], the optimized flow rate is 11.9839 kmol/hr with 10 stages.

TABLE 6.5

The Effect of IL Price on Optimization Results and Total Cost (Equation 6.33)

IL Cost Reduction	Total Cost (Crore INR/year)	N	*Sol* (kmol/hr)	IL Loss in Raffinate (kmol/hr)	Butanol Loss in Raffinate (kmol/hr)	Ethanol Loss in Raffinate (kmol/hr)
[TDTHP][DCA] (1)–Ethanol (2)–1-Butanol (3)–Water (4)						
1/1	11253.1	3	3.9891	0.1102	0.1183	0.0001
1/10	1129.8	2	4.6987	0.1102	0.1138	0.0003
1/100	113.7	9	20.3997	0.1114	0.0001	0
1/1000	11.4	5	40.0000	0.1114	0.0001	0
1/10000	1.2	5	40.0000	0.1114	0.0001	0
[TDTHP][Phosph] (1)–Ethanol (2)–1-Butanol (3)–Water (4)						
1/1	4526	10	36.2850	0.0551	0.0375	0.0321
1/10	455.8	10	30.5992	0.0551	0.0207	0.0316
1/100	47.8	10	11.9839	0.0555	0.0040	0.0251
1/1000	6.8	10	9.5960	0.0556	0.00563	0.023333
1/10000	2.6	10	9.4108	0.0556	0.0059	0.0232

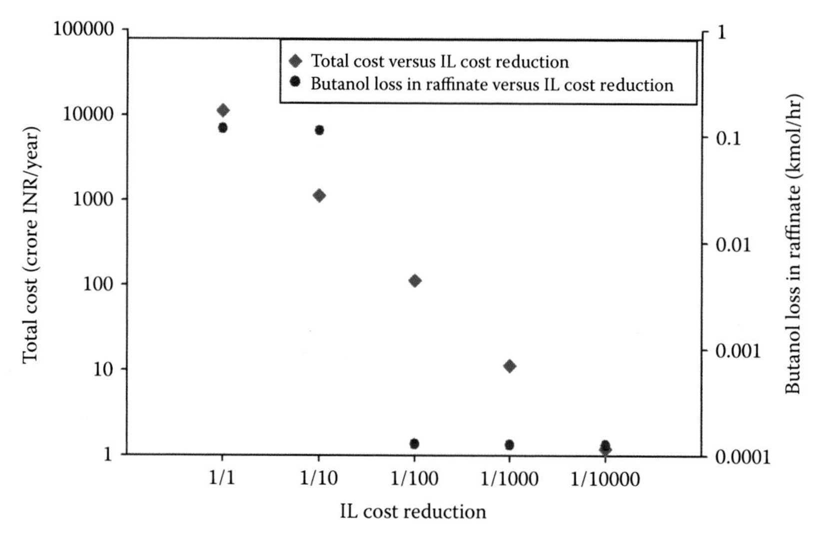

FIGURE 6.18
The effect of the IL cost on the total cost and butanol loss in raffinate for the [TDTHP][DCA] (1)–ethanol (2)–1-butanol (3)–water (4) system at $T = 298.15$ K and $p = 1$ atm.

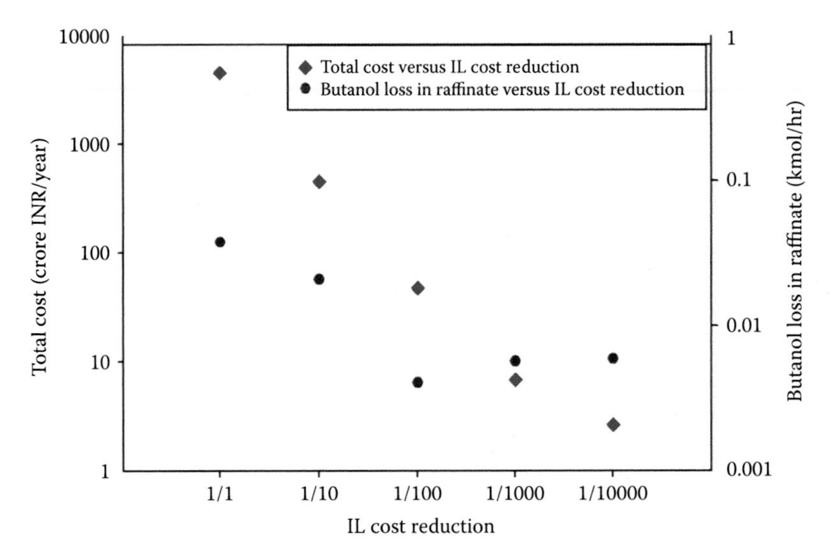

FIGURE 6.19
The effect of the IL cost on the total cost and butanol loss in raffinate for the [TDTHP][Phosph] (1)–ethanol (2)–1-butanol (3)–water (4) system at $T = 298.15$ K and $p = 1$ atm.

References

Arumugam, M. S., & Rao, M. V. C. (2008). On the improved performances of the particle swarm optimization algorithms with adaptive parameters, cross-over operators and root mean square (RMS) variants for computing optimal control of a class of hybrid systems. *Applied Soft Computing, 8*(1), 324–336. doi:10.1016/j.asoc.2007.01.010.

Colorni, A., Dorigo, M., & Maniezzo, V. (1991). Distributed optimization by ant colonies. In *Proceedings of the First European conference on Artificial Life* (pp. 134–142). Paris, France.

Dorigo, M. (1992). *Optimization, Learning and Natural Algorithms* (in Italian) (PhD thesis). Milan, Italy: Dipartimento di Elettronica, Politecnico di Milano.

Eberhart, R., & Kennedy, J. (1995a). A new optimizer using particle swarm theory. In *Proceedings of the Sixth International Symposium on Micro Machine and Human Science*, 39–43.

Eberhart, R., & Kennedy, J. (1995b). Particle swarm optimization. In *Proceedings of IEEE International Conference on Neural Networks, 4*, 1942–1948.

Elbeltagi, E., Hegazy, T., & Grierson, D. (2005). Comparison among five evolutionary-based optimization algorithms. *Advanced Engineering Informatics, 19*(1), 43–53. doi:10.1016/j.aei.2005.01.004.

Ethni, S. A., Zahawi, B., Giaouris, D., & Acarnley, P. P. (2009). Comparison of particle swarm and simulated annealing algorithms for induction motor fault identification. In *Proceedings of 7th IEEE International Conference on Industrial Informatics* (pp. 470–474). Wales, UK.

Goldberg, D. E. (1989). Genetic Algorithms in Search, Optimization and Machine Learning. Boston: Addison-Wesley Longman Publishing Co., Inc.

Hassan, R., Cohanim, B. E., & de Weck, O. L. (2005). A comparison of particle swarm optimization and the genetic algorithm. In *46th AIAA/ASME/ASCE/AHS/ASC Structures, Structural Dynamics, and Materials Conference*, number AIAA-2005-1897. Austin, TX: American Institute of Aeronautics and Astronautics.

Kirkpatrick, S., Gelatt, C. D., & Vecchi, M. P. (1983). Optimization by simulated annealing. *Science, 220*(4598), 671–680. doi:10.1126/science.220.4598.671.

Kitayama, S., Arakawa, M., & Yamazaki, K. (2006). Penalty function approach for the mixed discrete nonlinear problems by particle swarm optimization. *Structural and Multidisciplinary Optimization, 32*(3), 191–202. doi:10.1007/s00158-006-0021-2.

Shin, D. K., Gürdal, Z., & Griffin, O. H. (1990). A penalty approach for nonlinear optimization with discrete design variables. *Engineering Optimization, 16*(1), 29–42. doi:10.1080/03052159008941163.

Sivanandam, S. N., & Deepa, S. N. (2009). A comparative study using genetic algorithm and particle swarm optimization for lower order system modelling. *International Journal of the Computer, the Internet and Management, 17*(3), 1–10.

Storn, R., & Price, K. (1997). Differential evolution – A simple and efficient heuristic for global optimization over continuous spaces. *Journal of Global Optimization, 11*(4), 341–359. doi:10.1023/a:1008202821328.

Tsuboka, T., & Katayama, T. (1976). General design algorithm based on pseudo-equilibrium concept for multistage multi-component liquid-liquid separation processes. *Journal of Chemical Engineering of Japan, 9*(1), 40–45. doi:10.1252/jcej.9.40.

Ubaidullah, S., Upadhyay, R., Raut, S. A., & Rahman, I. (2013). Optimization of continuous extraction column and solvent selection using differential evolution technique. *Journal of the Institution of Engineers (India): Series E, 93*(1), 49–54. doi:10.1007/s40034-012-0004-3.

Zelinka, I. (2004). SOMA - Self-Organizing Migrating Algorithm. In New Optimization Techniques in Engineering, 141 of Studies in Fuzziness and Soft Computing, 167–217. Springer Berlin Heidelberg.

7

Cuckoo Search Optimization and Application to Liquid–Liquid Equilibrium

7.1 Introduction*

This chapter uses a new variant of the optimization technique, namely Cuckoo Search (CS) algorithm, to generate the liquid–liquid equilibrium (LLE) data. Liquid–liquid extraction is an important separation technology with a wide range of applications in chemical, petrochemical and pharmaceutical industries. The LLE data of multi-component systems are essential for proper understanding of the extraction process and for the designing and optimization of separation processes. Excess Gibbs free energy models, such as the nonrandom two-liquid (NRTL; Renon & Prausnitz, 1968) and the UNIversal QUAsiChemical (UNIQUAC; Abrams & Prausnitz, 1975) models, are commonly used to predict the LLE as they provide good agreement with experimental data (Banerjee, Singh, Sahoo, & Khanna, 2005; Santiago, Santos, & Aznar, 2009; Vatani, Asghari, & Vakili-Nezhaad, 2012). For LLE prediction, each of these models requires binary interaction parameters. These parameters are generally estimated from the known experimental LLE data by the optimization of an objective function. Mathematically, the aim of optimization is to find the set of inputs that either maximizes or minimizes the output of the objective function. In LLE modelling, the objective function is nonlinear and highly nonconvex having multiple local optima which makes most conventional methods (deterministic algorithms) inefficient and stuck in the wrong solutions. For the correct LLE prediction in liquid–liquid phase equilibria, finding the global optimum (reliable interaction parameters) thus becomes a necessary requirement.

Heuristic and metaheuristic algorithms are designed to deal with highly nonlinear and nonconvex problems. Most of these algorithms are

* Sections 7.1 and 7.2 reprinted (adapted) with permission from A. Bharti, Prerna, T. Banerjee, Applicability of Cuckoo Search algorithm for the prediction of multicomponent liquid–liquid equilibria for imidazolium- and phosphonium-based ionic liquids, *Ind. Eng. Chem. Res.* 54, 12393–12407, 2015. Copyright 2015 American Chemical Society.

nature-inspired or bio-inspired. Nature-inspired metaheuristic algorithms are becoming increasingly popular to solve global optimization problems. All stochastic algorithms with randomization and local search can be termed as metaheuristic algorithms. The two major components of any metaheuristic algorithm are: selection of the best solutions which ensures that the solutions will converge to the optimal and randomization which avoids the solutions being trapped at the local optima. Their efficiency is remarkable; these have many advantages over traditional, deterministic methods and thus have been applied in almost all areas of science, engineering and industry (Yang & Deb, 2014). Recent literature report many established nature-inspired metaheuristics, which are listed in Table 7.1. These algorithms are broadly classified into evolutionary algorithms, physical algorithms, swarm intelligence, bio-inspired algorithms and others (Nanda & Panda, 2014).

Genetic algorithm (GA), developed by John Holland and his collaborators, is based on Charles Darwin's theory of natural selection and remains one of the most widely used optimization algorithms in modern nonlinear optimization. Many variants of the GA have been developed and applied to a wide range of optimization problems. The GA starts with a set of randomly generated solutions called population. Then, the fitness of all the individuals in the population is evaluated and a new population is created by performing selection, crossover and mutation operations. The algorithm continues to evolve the population till certain stopping criteria are met (Yang, 2010).

Simulated annealing is a trajectory-based search algorithm which mimics the annealing process in material processing. Annealing is a process in metallurgy where metals are heated to specific high temperature and then slowly cooled to decrease defects and make them reach a state of low energy where they are very strong. After each iteration of the simulated annealing algorithm, a new solution is randomly generated. The algorithm accepts not only

TABLE 7.1

Broad Classification of Nature Inspired Metaheuristic Algorithms

Types	
Evolutionary algorithms	Genetic algorithm (GA) Differential evolution (DE) Genetic programming (GP)
Physical algorithms	Simulated annealing (SA) Harmony search (HS)
Swarm intelligence	Particle swarm optimization (PSO) Ant colony optimization (ACO) Artificial bee colony (ABC)
Bio-inspired algorithms	Artificial immune system (AIS) Krill herd algorithm
Other nature-inspired algorithms	Firefly algorithm Cuckoo Search algorithm Bat algorithm

all new solutions that lower the objective function but also, with a certain probability, the solutions that raise the objective function. By accepting solutions that raise the objective, the algorithm avoids being trapped in local minima in early iterations and is able to explore globally for better solutions (Yang, 2010). Particle swarm optimization (PSO) is inspired by the swarm intelligence of fish and birds which has been discussed in detailed in Chapter 6.

The application of nature-inspired metaheuristic algorithms for solving phase equilibrium calculations (PEC), phase stability (PS) problems and for parameter estimation (PE) has grown considerably in recent years (Bonilla-Petriciolet, Rangaiah, & Segovia-Hernández, 2010; Bonilla-Petriciolet & Segovia-Hernández, 2010; Fateen, Bonilla-Petriciolet, & Rangaiah, 2012; Fernández-Vargas, Bonilla-Petriciolet, & Segovia-Hernández, 2013; Srinivas & Rangaiah, 2007; Zhang, Fernández-Vargas, Rangaiah, Bonilla-Petriciolet, & Segovia-Hernández, 2011; Zhang, Rangaiah, & Bonilla-Petriciolet, 2011). Apart from the PEC and PS problems, stochastic global optimization methods have also been applied for the estimation of binary interaction parameters in multi-component LLE systems. Singh, Banerjee and Khanna (2005) utilized the GA to estimate the binary interaction parameters for the NRTL and the UNIQUAC models in multi-component LLE systems and demonstrated that their performance was better than the inside variance estimation method and the techniques applied in ASPEN and DECHEMA. Sahoo, Banerjee, Ahmad and Khanna (2006) calculated the interaction parameters for the NRTL model in ternary, quaternary and quinary LLE systems based on the GA and showed that the results obtained using the GA were better than other techniques in the literature. Ferrari, Nagatani, Corazza, Oliveira, and Corazza (2009) applied SA and PSO algorithms for PE of the NRTL and the UNIQUAC models for binary and multi-component LLE systems and showed that both algorithms were capable of modelling liquid–liquid equilibrium data. Merzougui, Hasseine, Kabouche, and Korichi (2011) used a hybrid algorithm, that is, a combination of the GA and the Levenberg–Marquardt (LM) method, for PE with the NRTL and the UNIQUAC models for six LLE systems and found good correlation with experimental data. Vatani et al. (2012) also performed the LLE calculation for 20 different ionic liquid (IL)-based ternary LLE systems by the NRTL and the two-suffix Margules models with binary interaction parameters calculated using the GA. A harmony search (HS) algorithm was used by Merzougui, Hasseine and Laiadi (2012) to calculate the interaction parameters of the NRTL model for 20 ternary liquid–liquid systems. Kabouche, Boultif, Abidi and Gherraf (2012) have used SA, GA, Nelder–Mead Simplex (NMS), SA-NMS (hybrid) and GA-NMS (hybrid) to estimate interactions parameters of the NRTL and the UNIQUAC models for LLE systems. The hybrid algorithm GA-NMS showed the best performance in terms of root-mean-square deviation (RMSD) with minimum number of iterations among the others. Bonilla-Petriciolet, Fateen and Rangaiah (2013) have recently analyzed the capabilities of seven stochastic global optimization methods, SA, GA, DE, PSO, HS, differential evolution with tabu list (DETL) and bare bones PSO (BBPSO), to model mean activity coefficients in aqueous solutions of quaternary ammonium salts at 25°C using the electrolyte NRTL model. The results indicated that SA, DETL and BBPSO offer better performance for solving PE problems involved in the modelling of the thermodynamic properties of ILs.

Besides the well-known methods, the investigations on nature-inspired optimization algorithms are still currently under development. CS is one of the latest nature-inspired metaheuristic algorithms developed by Yang and Deb (2009). It is a population-based method which mimics the breeding behavior such as brood parasitism of certain cuckoo species. This algorithm is enhanced by the so-called Lévy flights. Recent studies have shown that the CS algorithm is potentially far more efficient than other algorithms in many applications. It has been applied in many areas (Fister, Fister, & Fister, 2013) which include applied thermodynamic calculations. Bhargava, Fateen and Bonilla-Petriciolet (2013) have applied CS algorithm for solving PS, phase equilibrium and reactive phase equilibrium problems. They found that CS offers a reliable performance for solving these thermodynamic calculations and is better than other metaheuristics for phase equilibrium modelling. Fateen and Bonilla-Petriciolet (2014a) have compared the reliability and efficiency of eight promising nature-inspired metaheuristic algorithms for the solution of nine difficult PS and phase equilibrium problems. These algorithms are the Cuckoo Search (CS), intelligent firefly (IFA), bat (BA), artificial bee colony (ABC), MAKHA (a hybrid between monkey algorithm and krill herd algorithm), covariance matrix adaptation evolution strategy (CMAES), magnetic charged system search (MCSS) and BBPSO. The results clearly showed that CS is the most reliable of all tested optimization methods as it successfully solved all thermodynamic problems tested in the study. Fateen and Bonilla-Petriciolet (2014b) have also applied gradient-based cuckoo search (GBCS) algorithm for solving several challenging PS problems and analyzed its performance at different numerical effort levels. The GBCS was found to perform better than the original CS algorithm. In comparison with other stochastic optimization, GBCS proved to be the most reliable without any reduction in efficiency.

The computation using the CS algorithm is scarce for multi-component phase equilibria problems. In a recent work by Jaime-Leal, Bonilla-Petriciolet, Bhargava and Fateen (2015), the CS algorithm was used to predict the binary phase equilibrium data for aqueous quaternary ammonium IL mixtures. In their work, the mean activity coefficients of quaternary ammonium ILs were predicted using the e-NRTL model. The results obtained were very encouraging with a global success rate (SR) of ~86% with CS as compared to ~77% with other stochastic methods. However, the methods of computation with ternary, quaternary or quinary systems are not available in the literature. Keeping this limitation in mind, we have attempted to predict the multi-component LLE data for both IL and organic solvent systems. In our study, binary interaction parameters of the UNIQUAC and the NRTL models were estimated using the CS algorithm for 39 ternary systems which include 32 IL-based systems and 7 organic-solvent-based systems. The results (RMSD) thus obtained were compared with those reported in the literature. The performance of CS was compared with the GA and the PSO algorithms using three quaternary systems and one quinary system.

7.2 Cuckoo Search Algorithm

The CS algorithm is a natured-inspired metaheuristic search algorithm, based on the reproduction behavior of cuckoos, that has been recently developed by Yang and Deb, 2009; Yang and Deb 2013. Specific egg-laying and breeding behavior of some cuckoo species is the basis of this optimization algorithm. Cuckoo is a brood parasite, which never build their own nests. They lay their eggs in the nest of other species, leaving host species to care for the eggs. If these eggs are discovered by the host bird, it either throws out the cuckoo egg or abandons the nest to start afresh. Some species of cuckoos have learnt to mimic the colour and pattern of their own eggs so as to match that of their hosts. This reduces the probability of the eggs being abandoned and therefore increases their reproductivity. To implement these concepts, the CS algorithm employs three idealized rules:

1. Eggs are laid one at a time and then dumped randomly in a host nest.
2. The nest having better eggs (solutions) are carried over to next generation.
3. The host bird usually finds an alien egg within a probability of unity. However, if an alien egg is found, the host bird can either throw the egg away or abandon the nest and build a completely new nest.

The above rules are valid for single objective optimization problems. For multi-objective optimization problems with K different objectives, the first and the last rules need modification. Here, each cuckoo lays K eggs at a time (Rule 1) and thus a new nest with K eggs is built (Rule 3).

From implementation point of view, each egg in a nest represents a solution and a cuckoo egg represents a new solution. The aim is to use the potentially better solution (the cuckoo egg) to replace the not-so-good solution (the egg) in the nests.

The steps of the algorithm are as follows:

Step 1: Initialization of CS algorithm parameters

The following parameters are defined: number of nests (n), discovery probability (p_a), lower and upper bounds and maximum number of generations/iterations (iter_{max}).

Step 2: Generation of the initial population of nests of host birds

The initial population of the nests is generated randomly between the lower bound and the upper bound.

Step 3: Generation of new solutions/eggs by Lévy flights

In this step, all the host eggs except the best one are replaced by new cuckoo eggs if the objective function corresponding to each new cuckoo egg is better than the corresponding existing host egg objective function. A new solution is generated by Lévy flight as below

$$y_i^{(t+1)} = y_i^{(t)} + \alpha \cdot S \cdot \left(y_i^{(t)} - y_{\text{best}}^{(t)} \right) \cdot (\text{rand}) \tag{7.1}$$

where α is the step size parameter, 'rand' is a random number from a standard normal distribution, $y_i^{(t)}$ is the ith nest current position, y_{best}^{t} is the current best nest and S is a random walk based on the Lévy flights.

In the CS algorithm, Lévy flights are used to explore the unknown, large-scale search space as it is more efficient than the Brownian random walks. The Lévy flight essentially provides a random walk whose random step length is drawn from a Lévy distribution (Equation 7.2).

$$\text{Lévy} \approx u = t^{-1-\beta} \quad (1 < \beta \le 2) \tag{7.2}$$

This has an infinite variance with an infinite mean. In this work, a Lévy flight has been performed for generating a new solution according to the Mantegna algorithm (Yang, 2014). In Mantegna's algorithm, the step length S is calculated by

$$s = \frac{u}{|v|^{1/\beta}} \tag{7.3}$$

where β is considered to be 1.5 and u and v are drawn from normal distributions. It takes the following form:

$$u \approx N\left(0, \sigma_u^2\right), \qquad v \approx N\left(0, \sigma_v^2\right) \tag{7.4}$$

where

$$\sigma_u = \left\{ \frac{\Gamma(1+\beta)\sin\left(\pi\beta/2\right)}{\Gamma\left[(1+\beta)/2\right]\beta 2^{(\beta-1)/2}} \right\}^{1/\beta}, \qquad \sigma_v = 1 \tag{7.5}$$

Here, Γ is the standard Gamma function.

Step 4: Discovery of alien eggs

In this step, alien eggs' discovery is performed for each egg by generating a random number rand $\in [0,1]$ and comparing it with discovery probability p_a as follows:

$$\begin{aligned} y_i^{(t+1)} &\leftarrow y_i^{(t)}, & \text{if rand} \leq p_a \\ y_i^{(t+1)} &\leftarrow y_i^{(t)} + \zeta\left(y_j^t - y_k^t\right), & \text{if rand} > p_a \end{aligned} \tag{7.6}$$

where y_j^t and y_k^t are two different solutions selected randomly by random permutation and ζ is a random number drawn from a uniform distribution.

Step 5: Termination criterion

The maximum number of iterations or tolerance criteria may be used as the termination criterion of the algorithm. In this work, the maximum number of iterations has been used as a termination criterion. The pseudo-code and the flow diagram of the CS algorithm are shown in Figures 7.1 and 7.2, respectively.

7.3 Evaluation of Cuckoo Search, Genetic Algorithm and Particle Swarm Optimization Algorithms on Benchmark Functions

Four classical benchmark functions namely Ackley (function f_1), Rosenbrock (function f_2), Rastrigin (function f_3) and Griewank (function f_4; Table 7.2) were selected to compare the performance of the CS algorithm with the GA and the PSO algorithm. The description of the benchmark functions used in this work is listed in Table 7.2. The Rosenbrock function (function f_2) is uni-modal, having only one minimum. The others are multi-modal, with a considerable number of local minima in the search space. All functions except f_2 have their global minimum at the origin. For Rosenbrock function (f_2), the global minimum is at (1,1,…,1). Maximum number of iterations for the dimensions of 10, 20 and 30, were 500, 750 and 1000, respectively (Table 7.3). Other algorithm parameters are listed in Table 7.3. For each function, 20 trials were performed. The optimization algorithm has been coded in MATLAB® 2014b. The mean and the standard deviation obtained with the CS algorithm for each function are reported in Table 7.4 along with GA and PSO results which were taken from the literature (Karaboga & Basturk, 2007). CS outperformed GA and PSO algorithms while evaluating the function f_2 whereas underperformed for function f_3. In case of function f_1 and f_4, the performance of CS is equivalent to GA and PSO. Therefore, the CS algorithm has the ability to get out of the local minimum and can be efficiently used for the optimization of multi-variable and multi-modal functions.

```
begin
Objective function f(x), x = (x_1,...,x_d)^T
Generate initial population of
n host nests x_i (i = 1,2...,n)

while (t < MaxGeneration) or (stop criterion)
Get a cuckoo randomly by Levy flights
        evaluate its quality/fitness F_i
Choose a nest among n (say, j) randomly

if (F_i < F_j),
        replace j by the new solution;
end

A fraction (p_a) of worse nests
are abandoned and new ones are built;
Keep the best solutions
(or nests with quality solutions);
Rank the solutions and find the current best

end

Postprocess results and visualization
end
```

FIGURE 7.1
Pseudo-code of CS algorithm.

7.4 Activity Coefficient Models: Nonrandom Two-Liquid and UNIQUAC

In the UNIQUAC model, the activity coefficient, γ_i, of component 'i' in the multi-component system is given by

$$\ln\gamma_i = \ln\left(\frac{\Phi_i}{x_i}\right) + \frac{z}{2}q_i \ln\left(\frac{\theta_i}{\Phi_i}\right) + l_i - \frac{\Phi_i}{x_i}\sum_{j=1}^{c} x_j l_j + q\left(1 - \ln\sum_{j=1}^{c}\theta_j\tau_{ji} - \sum_{j=1}^{c}\frac{\theta_j\tau_{ij}}{\sum_{k=1}^{c}\theta_k\tau_{kj}}\right) \quad (7.7)$$

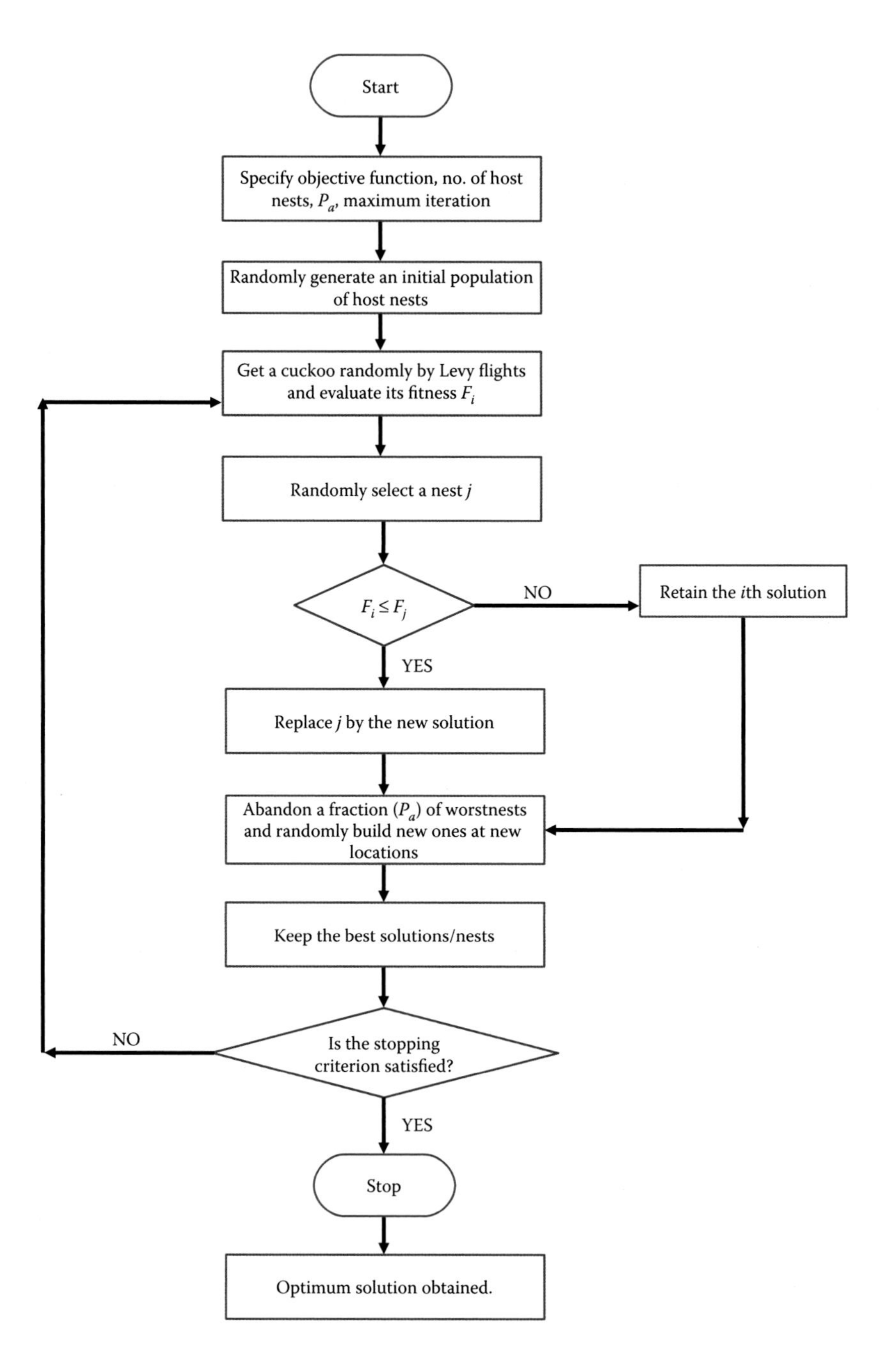

FIGURE 7.2
Flow diagram of CS algorithm.

TABLE 7.2

Benchmark Functions Used in This Study

Function	Mathematical Formulation	Search Space	Global Minimum
Ackley	$f_1(x) = -20\exp\left[-0.2\sqrt{\frac{1}{d}\sum_{i=1}^{d} x_i^2}\right] - \exp\left[\frac{1}{d}\sum_{i=1}^{d}\cos(2\pi x_i)\right] + (20+e)$	[−32.768,32.768]	0
Rosenbrock	$f_2(x) = \sum_{i=1}^{d-1}\left[(1-x_i)^2 + 100\left(x_{i+1} - x_i^2\right)^2\right]$	[−15,15]	0
Rastrigin	$f_3(x) = 10d + \sum_{i=1}^{d}\left[x_i^2 - 10\cos(2\pi x_i)\right]$	[−15,15]	0
Griewank	$f_4(x) = \frac{1}{4000}\sum_{i=1}^{d} x_i^2 - \prod_{i=1}^{d}\cos\left(\frac{x_i}{\sqrt{i}}\right) + 1$	[−600,600]	0

Source: Karaboga, D., and Basturk, B., *J. Global Optim.*, 39, 459–471, 2007.

TABLE 7.3

Algorithm Parameters

Algorithms	Number of Dimension	Number of Generations	Population Size	Discovery Probability (p_a)
GA	10	500[a]	125[a]	–
	20	750[a]	125[a]	–
	30	1000[a]	125[a]	–
PSO	10	500[a]	125[a]	–
	20	750[a]	125[a]	–
	30	1000[a]	125[a]	–
CS	10	500	20	0.25
	20	750	20	0.25
	30	1000	20	0.25

[a] Data from Karaboga, D. and Basturk, B., *J. Global Optim.*, 39, 459–471, 2007.

with

$$\tau_{ij} = \exp\left(-\frac{A_{ij}}{T}\right), \theta_i = \frac{q_i x_i}{\sum_k q_k x_k}, \Phi_i = \frac{r_i x_i}{\sum_k r_k x_k}, l_i = \frac{z}{2}(r_k - q_k) + 1 - r_k$$

TABLE 7.4

Results Obtained by GA, PSO and CS Algorithms

F	Alg. Dimension	GA[a]		PSO[a]		CS [This work]	
		Mean	SD	Mean	SD	Mean	SD
f_1	10	0.59267	0.22482	9.8499E-13	9.6202E-13	0.0252	0.0272
	20	0.92413	0.22599	1.1778E-6	1.5842E-6	0.4644	0.5254
	30	1.0989	0.24956	1.4917E-6	1.8612E-6	1.8558	0.7354
f_2	10	46.3184	33.8217	4.3713	2.3811	4.7789	1.9085
	20	103.93	29.505	77.382	94.901	16.3339	2.7356
	30	166.283	59.5102	402.54	633.65	57.9420	24.8518
f_3	10	1.3928	0.76319	2.6559	1.3896	13.4810	4.7705
	20	6.0309	1.4537	12.059	3.3216	50.9454	9.8580
	30	10.4388	2.6386	32.476	6.9521	96.6552	15.9904
f_4	10	0.050228	0.029523	0.079393	0.033451	0.0752	0.0262
	20	1.0139	0.026966	0.030565	0.025419	0.0634	0.0388
	30	1.2342	0.11045	0.011151	0.014209	0.1210	0.0811

[a] Data from Karaboga, D. and Basturk, B., *J. Global Optim.*, 39, 459–471, 2007.

where z is lattice coordination number ($z = 10$), r_i and q_i are, respectively, the volume and the surface area of the pure component i, x_i is the mole fraction of component i and τ_{ij} is interaction parameter between components i and j.

The structure parameters 'r' and 'q' for the compounds have been taken from our previous work (Banerjee et al., 2005; Bharti & Banerjee, 2015; Rabari & Banerjee, 2013, 2014) and the literature (Feng, Yang, Dang, & Wei, 2015; González, Calvar, González, & Domínguez, 2010; Santiago et al., 2009). In the NRTL model, the activity coefficient, γ_i, of component 'i' in the ternary system is given by

$$\ln \gamma_i = \frac{\sum_{j=1}^{c} \tau_{ji} G_{ji} x_j}{\sum_{l=1}^{c} G_{li} x_l} + \sum_{j=1}^{c} \frac{x_j G_{ij}}{\sum_{l=1}^{c} G_{lj} x_l} \left[\tau_{ij} - \frac{\sum_{r=1}^{c} x_r \tau_{rj} G_{rj}}{\sum_{l=1}^{c} G_{lj} x_l} \right] \tag{7.8}$$

with $G_{ji} = \exp(-\alpha_{ji}\tau_{ji})$, $\tau_{ji} = g_{ji} - g_{ii}/RT$, $\alpha_{ji} = \alpha_{ij}$

where g is an energy parameter characterizing the interaction of species i and j, x_i is the mole fraction of component i and α the non-randomness parameter. Although α can be treated as an adjustable parameter, in this study α was set equal to 0.2 which is a value prescribed for hydrocarbons.

7.5 Liquid–Liquid Equilibria Modelling

The thermodynamic equilibrium condition for multi-component liquid–liquid systems can be described by the following expression:

$$\gamma_i^I x_i^I = \gamma_i^{II} x_i^{II} \quad (i = 1, 2, 3, \ldots.) \tag{7.9}$$

where γ_i, the activity coefficient of component i in a phase (I or II), is predicted using the NRTL model. x_i^I and x_i^{II} represent the mole fraction of component i in phase I and II, respectively.

The compositions of the extract and the raffinate phases are calculated using a flash algorithm (Figure 7.3) as described by the modified Rachford–Rice algorithm (Seader & Henley, 2006). Optimum binary interaction parameters are those which minimize the difference between the experimental and calculated compositions and are given by the relation

$$F_{\mathrm{obj}} = \sum_{k=1}^{m} \sum_{i=1}^{c} \sum_{l=I}^{II} \left(x_{ik}^{l} - \hat{x}_{ik}^{l} \right)^2 \tag{7.10}$$

RMSD values, which provide a measure of the accuracy of the correlations, were calculated according to the following expression:

$$\mathrm{RMSD} = \left(\frac{F_{\mathrm{obj}}}{2mc} \right)^{1/2} = \left[\sum_{k=1}^{m} \sum_{i=1}^{c} \sum_{l=I}^{II} \frac{\left(x_{ik}^{l} - \hat{x}_{ik}^{l} \right)^2}{2mc} \right]^{1/2} \tag{7.11}$$

where m refers to the number of tie lines and c refers to the number of components. Here, x_{ik}^{l} and $\hat{x}_{ik}^{l}$ are, respectively, the experimental and predicted values of mole fraction for component i for the kth tie line in phase l. Figure 7.3 shows the flow diagram of the total algorithm used in this work for the calculation of binary interaction parameters.

7.6 Implementation of Cuckoo Search Algorithm for Multi-Component IL-Based Systems*

Thirty two IL and seven organic-based ternary systems were selected from the literature to test our approach with CS. They were correlated with both UNIQUAC and NRTL models with interaction parameters estimated by the CS algorithm. The systems used in this work are given in Tables 7.5 and 7.6,

* Sections 7.6 and 7.7 reprinted (adapted) with permission from A. Bharti, Prerna, T. Banerjee, Applicability of Cuckoo Search algorithm for the prediction of multicomponent liquid–liquid equilibria for imidazolium- and phosphonium-based ionic liquids, *Ind. Eng. Chem.* Res. 54, 12393–12407, 2015. Copyright 2015 American Chemical Society.

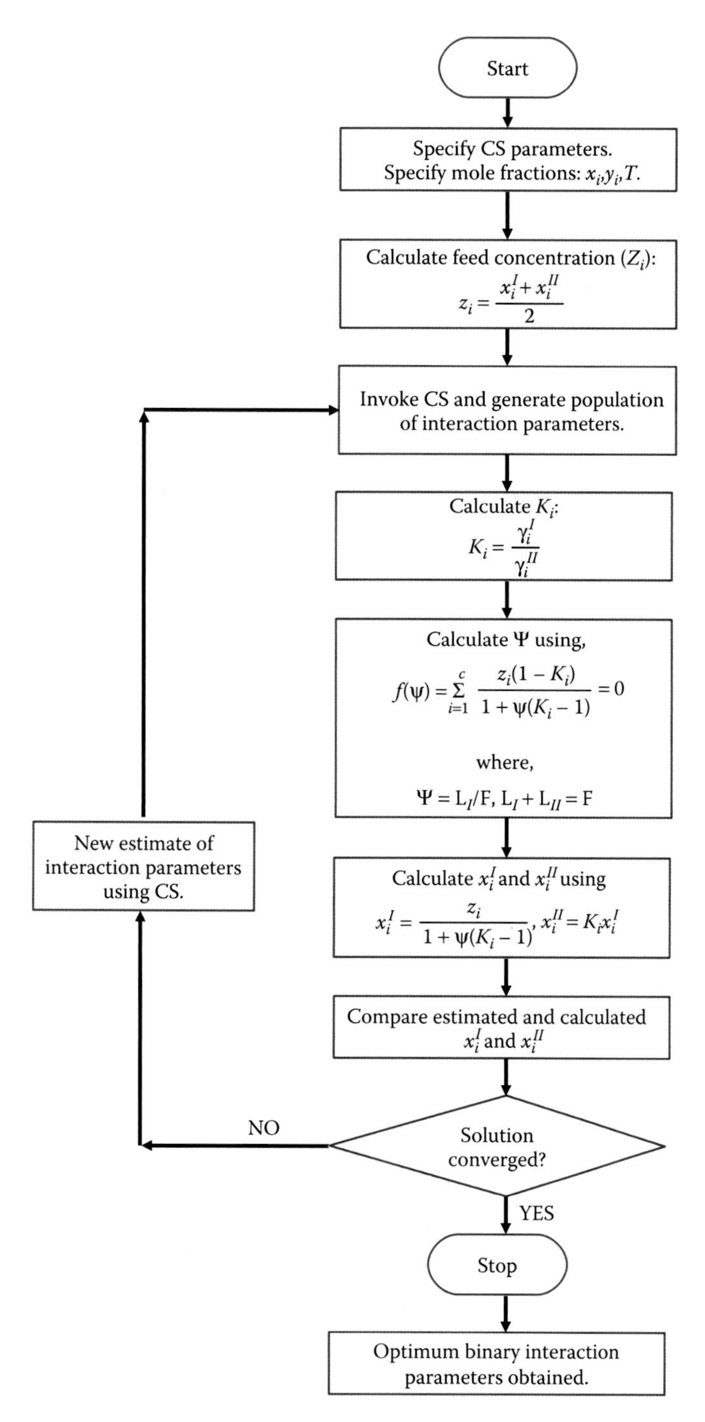

FIGURE 7.3
Flow diagram of the flash algorithm used for LLE modelling.

TABLE 7.5

Ionic Liquid-Based Ternary Systems Studied in This Work

System No.	System Name	Temperature (K)	Tie Line	Reference
1	[OMIM][Cl] + ethanol + TAEE	298.15	11	(Arce, Rodríguez, & Soto, 2004a)
2	[BMIM][TfO] + ethanol + TAEE	298.15	9	(Arce, Rodríguez, & Soto, 2004b)
3	[BMIM][TfO] + ethanol + ETBE	298.15	8	(Arce, Rodríguez, & Soto, 2006b)
4	[EMIM][TfO] + ethanol + ETBE	298.15	9	(Arce, Rodríguez, & Soto, 2006a)
5	[C_8MIM][BF_4] + thiophene + i-octane	298.15	11	(Alonso, Arce, Francisco, Rodríguez, & Soto, 2007)
6	[C_8MIM][BF_4] + cyclohexane + hexane	298.15	10	(Alonso et al., 2007)
7	[C_8MIM][BF_4] + thiophene + n-heptane	298.15	10	(Alonso, Arce, Francisco, & Soto, 2008)
8	[C_8MIM][BF_4] + thiophene + n-dodecane	298.15	12	(Alonso et al., 2008)
9	[C_8MIM][BF_4] + thiophene + n-hexadecane	298.15	11	(Alonso et al., 2008)
10	[HMIM][BF_4] + benzene + heptane	298.15	12	(Letcher & Reddy, 2005)
11	[HMIM][BF_4] + benzene + hexadecane	298.15	9	(Letcher & Reddy, 2005)
12	[HMIM][PF_6] + benzene + heptane	298.15	12	(Letcher & Reddy, 2005)
13	[OMIM][Cl] + benzene + heptane	298.15	5	(Letcher & Deenadayalu, 2003)
14	[HMIM][BF_4] + ethanol + hexene	298.15	12	(Letcher & Reddy, 2004)
15	[HMIM][BF_4] + ethanol + heptene	298.15	8	(Letcher & Reddy, 2004)
16	[HMIM][PF_6] + ethanol + hexene	298.15	13	(Letcher & Reddy, 2004)
17	[HMIM][PF_6] + ethanol + heptene	298.15	13	(Letcher & Reddy, 2004)
18	[HMIM][BF_4] + benzene + dodecane	298.15	7	(Letcher & Reddy, 2005)

(*Continued*)

TABLE 7.5 (*Continued*)

Ionic Liquid-Based Ternary Systems Studied in This Work

System No.	System Name	Temperature (K)	Tie Line	Reference
19	[HMIM][PF_6] + benzene + dodecane	298.15	7	(Letcher & Reddy, 2005)
20	[BMIM][BF_4] + benzene + heptane	298.15	8	(Revelli, Mutelet, & Jaubert, 2010)
21	[BMIM][BF_4] + thiophene + heptane	298.15	9	(Revelli et al., 2010)
22	[TDTHP][DCA] + 1-butanol + water	298.15	8	(Rabari & Banerjee, 2014)
23	[TDTHP][DEC] + 1-butanol + water	298.15	8	(Rabari & Banerjee, 2014)
24	[TDTHP][Phosph] + 1-propanol + water	298.15	8	(Rabari & Banerjee, 2013)
25	[TDTHP][Phosph] + 1-butanol + water	298.15	8	(Rabari & Banerjee, 2013)
26	[Emim][ESO_4] + benzene + hexane	298.15	8	(García, Fernández, Torrecilla, Oliet, & Rodríguez, 2009)
27	[Emim][ESO_4] + benzene + cyclohexane	298.15	9	(González et al., 2010)
28	[Emim][ESO_4] + benzene + methylcyclohexane	298.15	10	(González et al., 2010)
29	[Emim][ESO_4] + benzene + cyclooctane	298.15	9	(González et al., 2010)
30	[BMIM][Tf2N] + ethanol + water	283.20	13	(Cháfer, de la Torre, Font, & Lladosa, 2015)
31	[BMIM][Tf2N] + ethanol + water	303.20	10	(Cháfer et al., 2015)
32	[BMIM][Tf2N] + ethanol + water	323.20	8	(Cháfer et al., 2015)
		Total Tie Lines	**305**	

TABLE 7.6

Organic-Solvent-Based Ternary Systems Studied in This Work

System No.	System Name	Temperature (K)	Tie Line	Reference
1	Formamide + indole + 2-methylnaphthalene	308.15	12	(Feng et al., 2015)
2	Ethylene glycol + indole + 2-methylnaphthalene	308.15	12	(Feng et al., 2015)
3	Monoethanolamine + indole + 2-methylnaphthalene	308.15	10	(Feng et al., 2015)
4	2-methoxy-2-methylpropane + propionic acid + water	298.2	8	(Luo, Liu, Li, & Chen, 2015)
5	2-methoxy-2-methylpropane + propionic acid + water	323.2	8	(Luo et al., 2015)
6	2-methoxy-2-methylpropane + butyric acid + water	298.2	8	(Luo et al., 2015)
7	2-methoxy-2-methylpropane + butyric acid + water	323.2	8	(Luo et al., 2015)
		Total Tie Lines	**66**	

respectively. Structure parameters r and q for all compounds studied in this work have been reported in Tables 7.7 and 7.8. System 1, [OMIM][Cl] + Ethanol + TAEE, has been used for benchmarking study.

7.6.1 Effect of Bounds

The effect of the value of bounds on RMSD values has been considered and shown in Table 7.9 for the UNIQUAC model and Table 7.10 for the NRTL model for System 1. A population size of 20 was used for the benchmarking system. Maximum number of iterations, that is, 2000, was used as the stopping condition. Based on the recommendations of Yang and Deb (2009), p_a of 0.25 was used. For each bound, 30 numerical trials have been carried out with random initial values of interaction parameters. It can be inferred from Tables 7.9 and 7.10 that the RMSD values are minimum at a bound of −1000 to +1000 for the UNIQUAC model and −100 to +100 for the NRTL model. All the parameters were found to lie within the same range. Keeping this in mind, we have chosen the lower and the upper bounds for the estimation of interaction parameters for all the systems.

7.6.2 Maximum Number of Iterations and Population Size

The effect of the maximum number of iterations and the population size on RMSD values has been considered and shown in Figure 7.4a and b. For the UNIQUAC model, the population size of 20 and maximum iterations of 1000 are sufficient to give a very good RMSD value. A further increase in

TABLE 7.7

Abbreviation and Full Name of Ionic Liquids Used in This Work along with UNIQUAC Volume and Surface Area Structural Parameters

Abbreviation	Full Name of Ionic Liquid	r	q
[OMIM][Cl]	1-octyl-3-methylimidazolium chloride	11.993[a]	7.886[a]
[BMIM][TfO]	1-butyl-3-methylimidazolium trifluoromethanesulfonate	12.460[a]	7.518[a]
[OMIM][BF_4]	1-octyl-3-methylimidazolium tetrafluoroborate	13.187[a]	8.357[a]
[HMIM][BF_4]	1-hexyl-3-methylimidazolium tetrafluoroborate	11.658[a]	7.388[a]
[HMIM][PF_6]	1-hexyl-3-methylimidazolium hexafluorophosphate	12.869[a]	8.166[a]
[BMIM][BF_4]	1-Butyl-3-methylimidazolium tetrafluoroborate	10.057[a]	6.368[a]
[TDTHP][DCA]	Trihexyl(tetradecyl)-phosphonium dicyanamide	8.37[b]	5.81[b]
[TDTHP][DEC]	Trihexyl(tetradecyl)-phosphonium decanoate	8.77[b]	5.96[b]
[TDTHP][Phosph]	Trihexyl(tetradecyl)-phosphonium bis(2,4,4-trimethylpentyl) phosphinate	9.834[c]	6.258[c]
[BMIM][Tf2N]	1-butyl-3-methylimidazolium bis(trifluoromethylsulfonyl)imide	11.964[d]	9.753[d]
[EMIM][ESO_4]	1-ethyl-3-methylimidazolium Ethylsulfate	8.3927[e]	6.6260[e]
[EMIM][TfO]	1-ethyl-3-methylimidazolium trifluoromethanesulfonate	7.6193[f]	6.075[f]

[a] Data from Banerjee, T. et al., *Fluid Phase Equilib.*, 234, 64–76, 2005.
[b] Data from Rabari, D. and Banerjee, T. *Ind. Eng. Chem. Res.*, 53, 18935–18942, 2014.
[c] Data from Rabari, D. and Banerjee, T. *Fluid Phase Equilib.*, 355, 26–33, 2013.
[d] Data from Bharti, A. and Banerjee, T. *Fluid Phase Equilib.*, 400, 27–37, 2015.
[e] Data from Varma, N. R. et al., *Chem. Eng. J.*, 166, 30–39, 2011.
[f] Data from Santiago, R. S. et al., *Fluid Phase Equilib.*, 278, 54–61, 2009.

the population size and the number of iterations was not able to improve the solution (Figure 7.4a). For the NRTL model, the population size of 20 and maximum iterations of 1500 were able to give a very high RMSD value (Figure 7.4b). Keeping this in mind, we have chosen 20 as the population size for all systems and 1000 and 1500 as the maximum number of iterations, respectively, for the UNIQUAC and the NRTL models to estimate the interaction parameters in all systems. After the benchmarking with the above system, CS has been applied on the other ternary systems. For each system, 30 trials have been carried out with random initial values of population and the lowest RMSD along with the corresponding interaction parameters was then selected as the final result.

7.6.3 Comparison with Reported Data

The convergence plots of CS for few selected imidazolium- and phosphonium-based systems have been shown in Figure 7.5a–d. For imidazolium-based ILs (Figure 7.5a), the objective function value reduces to

TABLE 7.8

UNIQUAC Volume and Surface Area Structural Parameters of Compounds Used in This Work

Solvent	r	q
Tert-amyl ethyl ether (TAEE)	6.68	4.512
Ethanol	2.112	1.950
i-octane	5.986	4.642
Water	0.92	1.42
Hexane	4.594	3.621
n-heptane	5.292	4.152
Benzene	3.295	2.611
1-hexene	4.339	3.448
1-heptene	5.032	3.983
n-dodecane	8.5462[a]	7.096[a]
n-hexadecane	11.2438[a]	9.256[a]
Ethyl tert-butyl Ether (ETBE)	5.1726[a]	4.388[a]
Thiophene	2.8569[a]	2.140[a]
1-butanol	3.92[b]	3.67[b]
1-propanol	3.2499[b]	3.128[b]
Methylcyclohexane	4.72[c]	3.776[c]
Cyclooctane	5.395[c]	4.32[c]
Cyclohexane	4.0464[c]	3.24[c]
Formamide	1.6928[d]	1.644[d]
Indole	4.282[d]	2.692[d]
2-methylnaphthalene	5.7158[d]	4.008[d]
Ethylene glycol	2.4088[d]	2.248[d]
Monoethanolamine	2.5735[d]	2.36[d]
Propionic acid	2.8768[e]	2.612[e]
2-methoxy-2-methylpropane	4.0678[e]	3.632[e]
Butyric acid	3.5512[e]	3.152[e]

[a] Data from Santiago, R. S. et al., *Fluid Phase Equilib.*, 278, 54–61, 2009.
[b] Data from Rabari, D., and Banerjee, T. *Ind. Eng. Chem. Res.*, 53, 18935–18942, 2014.
[c] Data from González, E. J. et al., *J. Chem. Eng. Data*, 55, 4931–4936, 2010.
[d] Data from Feng, Y. et al., *Fluid Phase Equilib.*, 398, 10–14, 2015.
[e] Data from Luo, L. et al., *Fluid Phase Equilib.*, 403, 30–35, 2015.
Other data taken from Banerjee, T. et al. *Fluid Phase Equilib.*, 234, 64–76, 2005.

the order of 10^{-3} in approximately 60 iterations with the UNIQUAC model. For phosphonium ILs (Figure 7.5b) with the UNIQUAC model, the objective function value reduces to the order of 10^{-3} in approximately 100, 250 and 350 iterations, respectively, for [Phosph], [DCA] and [DEC] anions. With the NRTL model, imidazolium ILs (Figure 7.5c) used more iterations as compared to phosphonium ILs (Figure 7.5d) to obtain a function value of the order of 10^{-3}.

TABLE 7.9

Effect of Bounds on RMSD for System 1: [OMIM][Cl] + Ethanol + TAEE Using UNIQUAC Model

Bounds		RMSD			Hitting Bound	
Lower	**Upper**	**Min.**	**Avg.**	**Max.**	**Lower**	**Upper**
−100	+100	0.0336	0.0336	0.0336	Yes	Yes
−200	+200	0.0226	0.0226	0.0226	Yes	Yes
−500	+500	0.0058	0.0058	0.0058	No	Yes
−1000	+1000	0.0058	0.0058	0.0058	No	No
−1500	+1500	0.0058	0.0058	0.0058	No	Yes
−2000	+2000	0.0057	0.0058	0.0059	No	Yes

TABLE 7.10

Effect of Bounds on RMSD for System 1: [OMIM][Cl] + Ethanol + TAEE Using NRTL Model

Bounds		RMSD			Hitting Bound	
Lower	**Upper**	**Min.**	**Avg.**	**Max.**	**Lower**	**Upper**
−100	+100	0.0031	0.0047	0.0110	No	No
−200	+200	0.0033	0.0066	0.0146	No	Yes
−500	+500	0.0041	0.0108	0.0253	No	Yes
−1000	+1000	0.0041	0.0120	0.0213	No	Yes
−1500	+1500	0.0045	0.0144	0.0223	No	Yes
−2000	+2000	0.0052	0.0150	0.0223	No	Yes

Figure 7.6a–d shows the convergence plots of CS for imidazolium and phosphonium ILs with respect to the maximum number of iterations ($Iter_{max}$). For each $Iter_{max}$, 30 different trials have been carried out and the obtained minimum value of the objective function has been assigned as F_{obj}. F^*_{obj} is the global optimum calculated using CS with $Iter_{max} = 5000$. From the plot, it is clear that the performance of CS improves with the increment of $Iter_{max}$. In particular, CS is very reliable for finding the global solution with high precision for phosphonium ILs as compared to imidazolium ILs. This is due to the fact that the solution was within 10^{-7}–10^{-8} of the global minimum for phosphonium ILs with the UNIQUAC model. On the other hand, CS was able to find the global minimum with a tolerance of 10^{-6}–10^{-7} for imidazolium ILs with the UNIQUAC model. Again with the NRTL model, CS could converge to the global minimum within a tolerance of 10^{-3}–10^{-5} for imidazolium ILs and 10^{-6}–10^{-7} for phosphonium ILs.

For the calculation of the SR of CS, the following criterion has been considered: $F_{obj} - F^*_{obj} \leq \varepsilon$, where ε is tolerance. Tables 7.11 and 7.12 show the percentage SR (%SR) of CS with different tolerance values and stopping

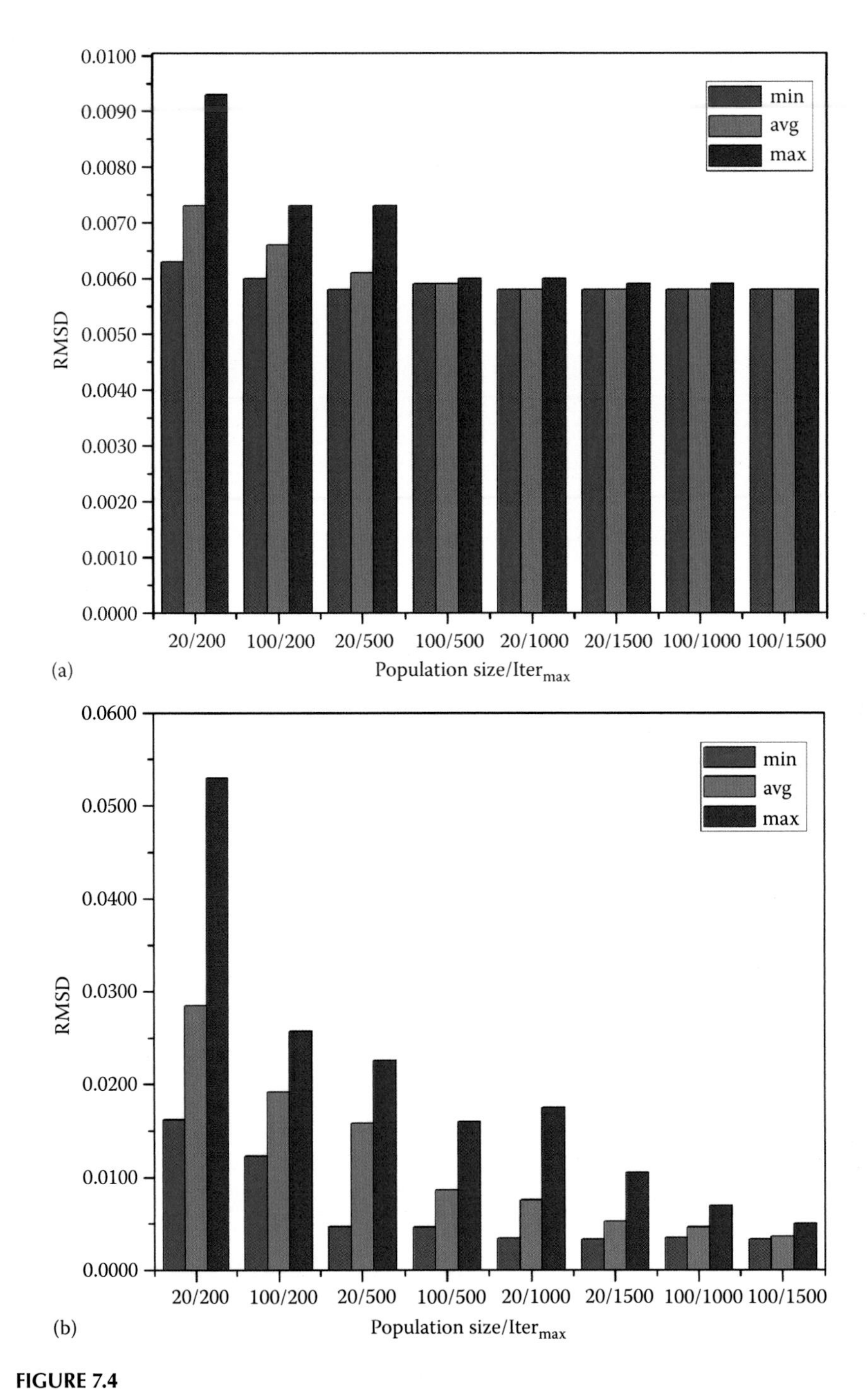

FIGURE 7.4
Effect of population size and $Iter_{max}$ on RMSD for System 1 (benchmarking system) using (a) the UNIQUAC model and (b) the NRTL model.

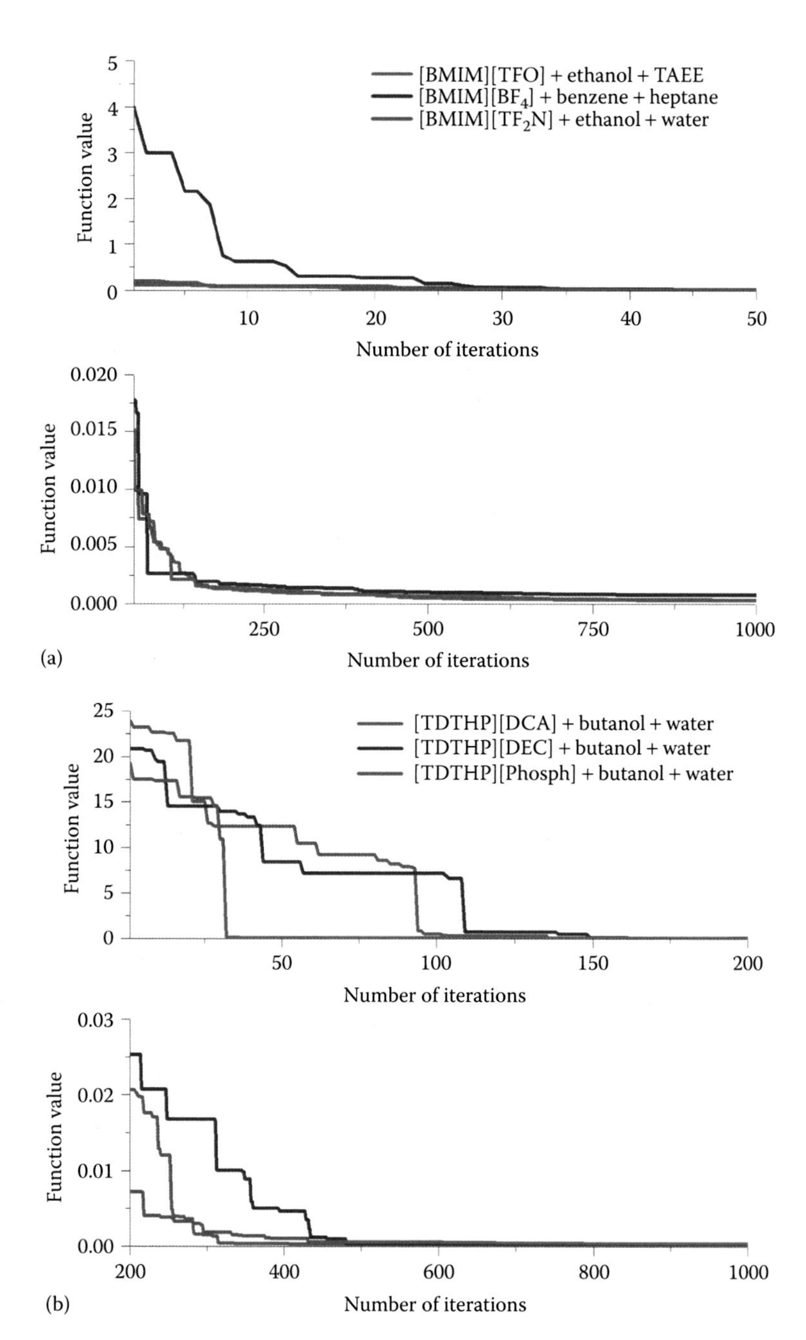

FIGURE 7.5
Convergence plots of CS for selected (a) imidazolium-based ternary systems with the UNIQUAC model (N = 20, Iter_{max} = 1000) and (b) phosphonium-based ternary systems with the UNIQUAC model (N = 20, Iter_{max} = 1000). *(Continued)*

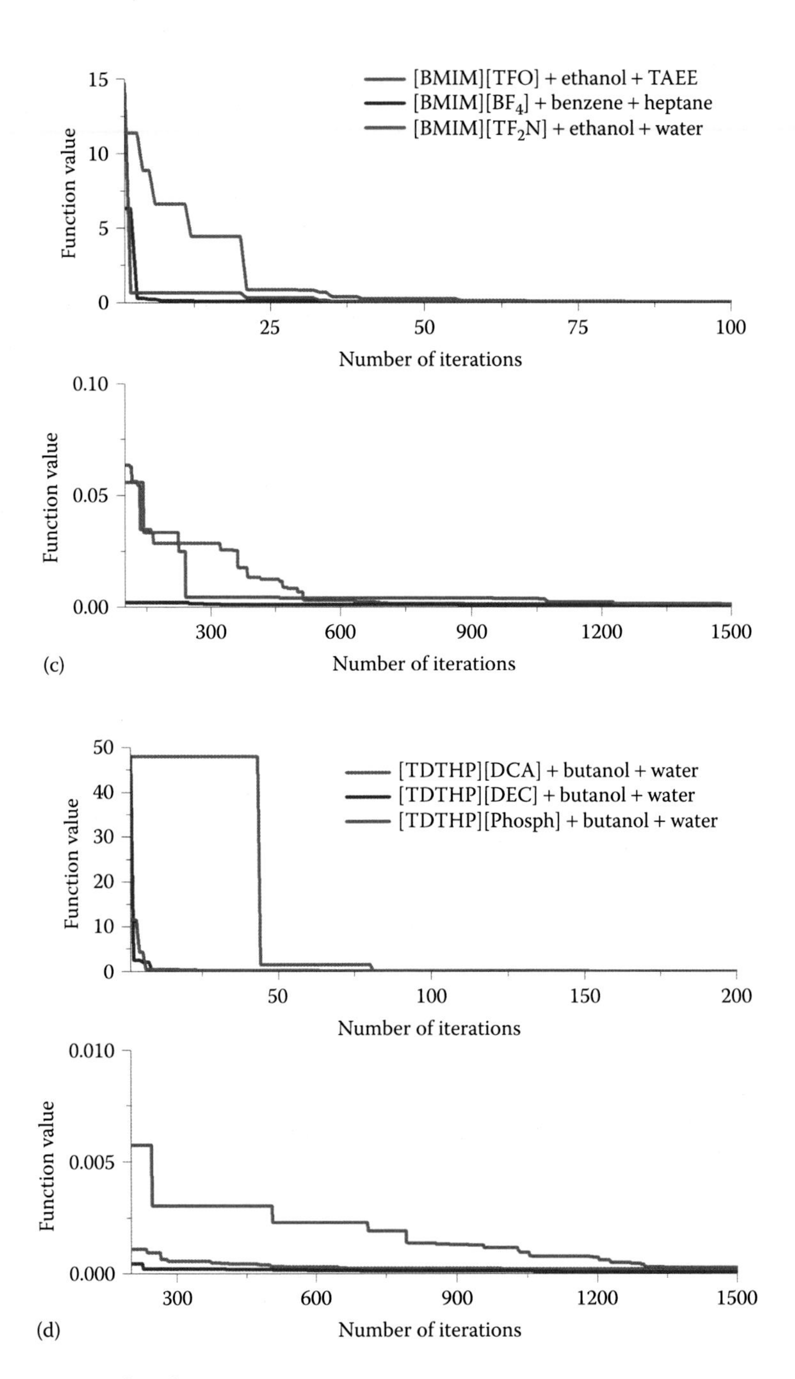

FIGURE 7.5 (Continued)
Convergence plots of CS for selected (c) imidazolium-based ternary systems with the NRTL model ($N = 20$, $Iter_{max} = 1500$) and (d) phosphonium-based ternary systems with the NRTL model ($N = 20$, $Iter_{max} = 1500$).

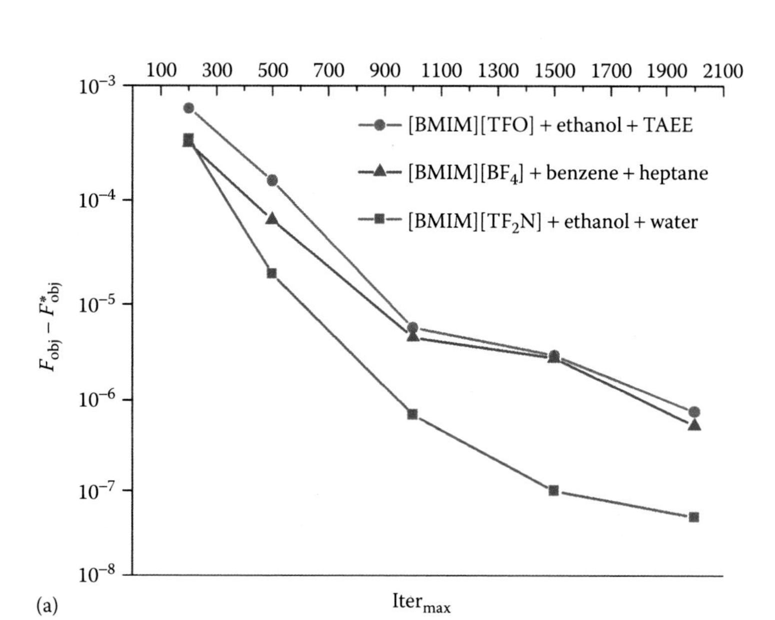

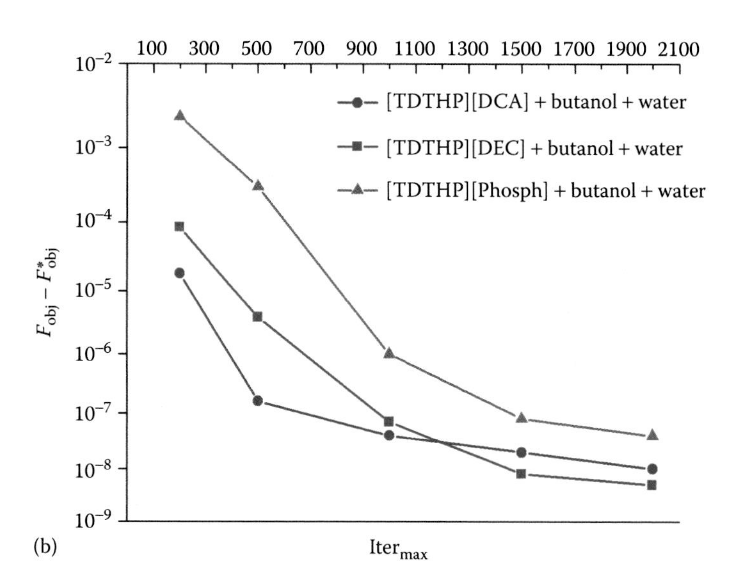

FIGURE 7.6
Convergence plots of CS for selected (a) imidazolium-based ternary systems (UNIQUAC model) w.r.t. $Iter_{max}$ and (b) phosphonium-based ternary systems (UNIQUAC model) w.r.t. $Iter_{max}$.

(*Continued*)

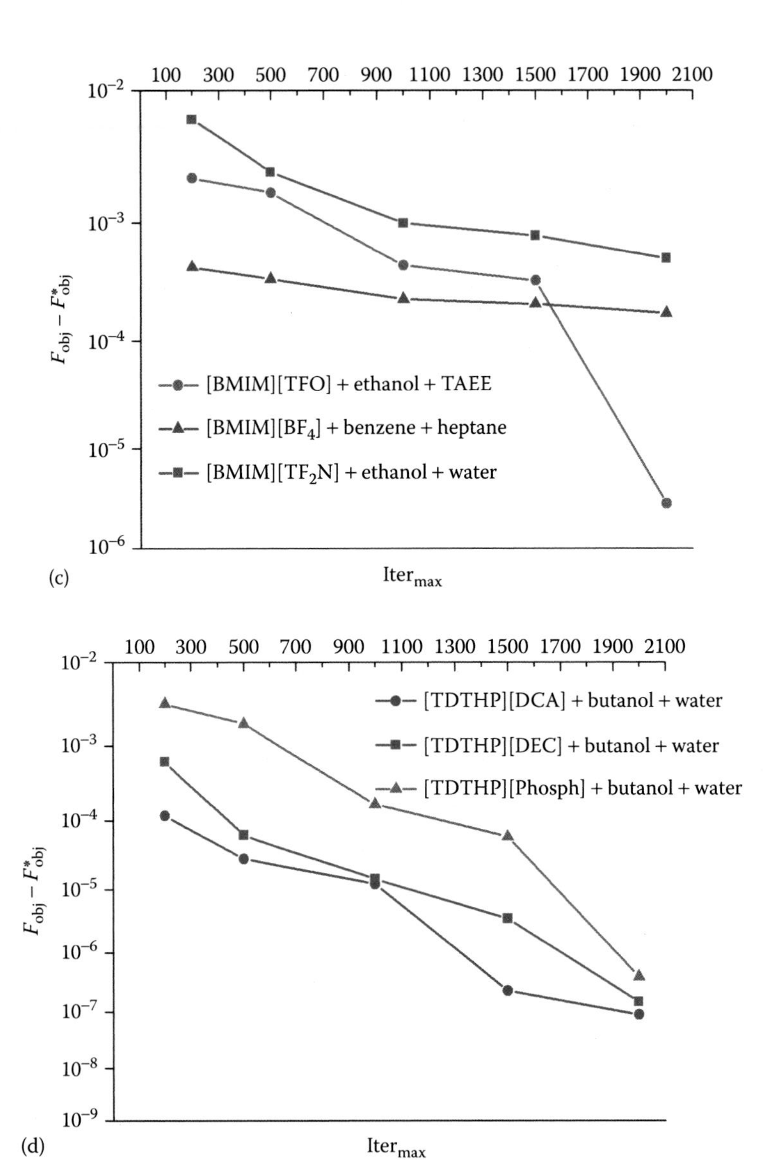

FIGURE 7.6 (Continued)
Convergence plots of CS for selected (c) imidazolium-based ternary systems (NRTL model) w.r.t. $Iter_{max}$ and (d) phosphonium-based ternary systems (NRTL model) w.r.t. $Iter_{max}$.

conditions. It is clear that %SR of CS increased with the increment of $Iter_{max}$ and decreased with the increment of tolerance values. With the UNIQUAC model, %SR for imidazolium ILs are in the range 23%–87% ($Iter_{max}$ = 2000), whereas for phosphonium ILs and ($Iter_{max}$ = 2000) at a tolerance value of 10^{-5}, >90% is obtained. With the NRTL model having maximum number

TABLE 7.11

Success Performance (%SR) of CS with the UNIQUAC Model for Selected Ternary Systems

Tolerance (ε)	$Iter_{max}$	SYS-2	SYS-20	SYS-30	SYS-22	SYS-23	SYS-25
1.00E-03	200	3.3	50.0	53.3	73.3	36.7	0.0
	500	76.7	93.3	100.0	100.0	93.3	33.3
	1000	100.0	100.0	100.0	100.0	100.0	96.7
	1500	100.0	100.0	100.0	100.0	100.0	100.0
	2000	100.0	100.0	100.0	100.0	100.0	100.0
1.00E-04	200	0.0	0.0	0.0	33.3	3.3	0.0
	500	0.0	6.7	23.3	100.0	36.7	0.0
	1000	26.7	40.0	63.3	100.0	93.3	50.0
	1500	26.7	76.7	83.3	100.0	96.7	86.7
	2000	40.0	96.7	93.3	100.0	96.7	96.7
1.00E-05	200	0.0	0.0	0.0	0.0	0.0	0.0
	500	0.0	0.0	0.0	53.3	6.7	0.0
	1000	3.3	6.7	23.3	90.0	76.7	26.7
	1500	3.3	23.3	60.0	96.7	96.7	76.7
	2000	23.3	86.7	70.0	96.7	96.7	90.0

Notes: SYS-2: [BMIM][TfO] + Ethanol + TAEE; SYS-20: [BMIM][BF4] + Benzene + Heptane; SYS-30: [BMIM][Tf_2N] + Ethanol + Water; SYS-22: [TDTHP][DCA] + Butanol + Water; SYS-23: [TDTHP][DEC] + Butanol + Water; SYS-25: [TDTHP][Phosph] + Butanol + Water.

of iterations as 2000, %SR for imidazolium ILs are in the range 0%–3.3%, whereas for phosphonium ILs they are in the range 3.3%–23.3% with a tolerance value of 10^{-5}. This implies that CS gave a lower order of success in the PE of imidazolium ILs, especially with the NRTL model. Thus, CS can be a recommended tool for imidazolium as well as phosphonium ILs with UNIQUAC model.

Tables 7.13 and 7.14 show the %SR of CS as a function of the cation- and anion-type ILs. With the UNIQUAC model, %SR of imidazolium IL with [BF_4] anion is greater than [PF_6] anion ($Iter_{max}$ = 2000) at a tolerance value of 10^{-5}, while for phosphonium ILs, the different anion types (DCA/DEC/Phosph) do not make much effect on %SR of CS. With same [TFO] anion, %SR of [EMIM] IL is greater than [BMIM] IL at a tolerance value of 10^{-5}.

The RMSD values calculated using the CS algorithm have been compared with the RMSD values reported in the literature and tabulated in Tables 7.15 through 7.18. For IL-based liquid–liquid ternary systems, the global RMSD values with CS are 0.0056 (Table 7.15) and 0.0076 (Table 7.17) for the UNIQUAC and the NRTL models, respectively. This is 64% and 45% better than the global values of 0.0156 and 0.0139 reported in the literature for 305 tie lines. For organic-solvent-based liquid–liquid ternary systems, the

TABLE 7.12

Success Performance (%SR) of CS with the NRTL Model for Selected Ternary Systems

ε	$Iter_{max}$	SYS-2	SYS-20	SYS-30	SYS-22	SYS-23	SYS-25
1.00E-03	200	0.0	16.7	0.0	86.7	13.3	0.0
	500	0.0	76.7	0.0	100.0	90.0	0.0
	1000	6.7	100.0	3.3	100.0	96.7	13.3
	1500	13.3	100.0	3.3	100.0	100.0	73.3
	2000	36.7	100.0	10.0	100.0	100.0	76.7
1.00E-04	200	0.0	0.0	0.0	0.0	0.0	0.0
	500	0.0	0.0	0.0	70.0	10.0	0.0
	1000	0.0	0.0	0.0	96.7	60.0	0.0
	1500	0.0	0.0	0.0	100.0	83.3	23.3
	2000	13.3	0.0	0.0	100.0	86.7	46.7
1.00E-05	200	0.0	0.0	0.0	0.0	0.0	0.0
	500	0.0	0.0	0.0	0.0	0.0	0.0
	1000	0.0	0.0	0.0	0.0	0.0	0.0
	1500	0.0	0.0	0.0	16.7	10.0	0.0
	2000	3.3	0.0	0.0	23.3	20.0	3.3

Notes: SYS-2: [BMIM][TfO] + Ethanol + TAEE; SYS-20: [BMIM][BF4] + Benzene + Heptane; SYS-30: [BMIM][Tf_2N] + Ethanol + Water; SYS-22: [TDTHP][DCA] + Butanol + Water; SYS-23: [TDTHP][DEC] + Butanol + Water; SYS-25: [TDTHP][Phosph] + Butanol + Water.

global RMSD values with CS are 0.0036 (Table 7.16) and 0.0048 (Table 7.18) for the UNIQUAC and the NRTL models, respectively. These again are 53% and 43% better than global value of 0.0077 and 0.0085 reported in the literature for 66 tie lines. The overall RMSD values with CS are 0.0053 and 0.0072 for the UNIQUAC and the NRTL models, respectively, for 371 tie lines. So in a nutshell, these are 63% and 45% better than the global values of 0.0145 and 0.0131 reported in the literature.

These results are extremely satisfactory when compared to the deviation reported by Santiago et al. (2009) and Aznar (2007). They are also about the same order of magnitude as reported by Vatani et al. (2012). Santiago et al. further correlated the LLE data of 50 ternary systems involving 12 different ILs, comprising 408 experimental tie lines, using the UNIQUAC model with a global deviation of 1.75%. Aznar correlated the LLE data for 24 IL-based ternary systems (184 tie lines) using the NRTL model and found global RMSD of 1.4%. Both Aznar (2007) and Santiago et al. (2009) estimated the interaction parameters using the Simplex method. Vatani et al. (2012) performed the LLE calculation for 20 different IL-based ternary systems using the NRTL model with binary interaction parameters calculated using the GA. The overall RMSD value for 169 tie lines was 0.39%.

TABLE 7.13

Effect of the Anion Type on the Success Performance (%SR) of CS with the UNIQUAC Model for Selected Ternary Systems

Tolerance (ε)	$Iter_{max}$	SYS-18	SYS-19	SYS-22	SYS-23	SYS-25
1.00E-03	200	3.3	0.0	73.3	36.7	0.0
	500	30.0	16.7	100.0	93.3	33.3
	1000	80.0	70.0	100.0	100.0	96.7
	1500	96.7	96.7	100.0	100.0	100.0
	2000	100.0	100.0	100.0	100.0	100.0
1.00E-04	200	0.0	0.0	33.3	3.3	0.0
	500	0.0	3.3	100.0	36.7	0.0
	1000	20.0	10.0	100.0	93.3	50.0
	1500	43.3	33.3	100.0	96.7	86.7
	2000	86.7	70.0	100.0	96.7	96.7
1.00E-05	200	0.0	0.0	0.0	0.0	0.0
	500	0.0	0.0	53.3	6.7	0.0
	1000	3.3	3.3	90.0	76.7	26.7
	1500	20.0	6.9	96.7	96.7	76.7
	2000	46.7	13.3	96.7	96.7	90.0

Notes: SYS-18: [HMIM][BF4] + Benzene + dodecane; SYS-19: [HMIM][PF6] + Benzene + dodecane; SYS-22: [TDTHP][DCA] + Butanol + Water; SYS-23: [TDTHP][DEC] + Butanol + Water; SYS-25: [TDTHP][Phosph] + Butanol + Water.

7.7 Comparison between Cuckoo Search, Genetic Algorithm and Particle Swarm Optimization Algorithms for Multi-Component Systems

In order to confirm our capability with CS, a comparison with GA and PSO algorithms was made for quaternary and quinary systems. Three quaternary systems, Ethanol + Water + Pentane + Hexane; Ethanol + Water + Pentane + Cyclohexane and Ethanol + Water + Hexane + Cyclohexane and one quinary system, Ethanol + Water + Pentane + Hexane + Cyclohexane, as studied by Khansary and Sani (2014) were selected for the same. The experimental data for the above-mentioned systems has been taken from Huang, Chung, Tseng and Lee (2010).

For each system, 20 trials have been carried out and the lowest RMSD along with the corresponding interaction parameters was selected as the final result and reported in Tables 7.19 and 7.20. For the quaternary system, CS gave RMSD (%) values in the range 0.14%–0.44% against ~1.0% for the GA and the PSO algorithm (Table 7.19), whereas for the quinary system, CS gave RMSD (%) values ~0.85% against ~2.0% for GA and PSO (Table 7.20). This clearly shows that the CS algorithm is a reliable tool to correlate the LLE data and in some cases, it is even superior to the GA and the PSO algorithm.

TABLE 7.14

Effect of the Cation Type on the Success Performance (%SR) of CS with the UNIQUAC Model for Selected Ternary Systems

Tolerance (ε)	$Iter_{max}$	SYS-3	SYS-4
1.00E-03	200	13.3	16.7
	500	66.7	90.0
	1000	100.0	96.7
	1500	100.0	96.7
	2000	100.0	100.0
1.00E-04	200	0.0	0.0
	500	0.0	6.7
	1000	6.7	46.7
	1500	53.3	86.7
	2000	73.3	100.0
1.00E-05	200	0.0	0.0
	500	0.0	0.0
	1000	0.0	10.0
	1500	16.7	53.3
	2000	30.0	90.0

Notes: SYS-3: [BMIM][TFO] + Ethanol + ETBE; SYS-4: [EMIM][TFO] + Ethanol + ETBE.

In summary, based on %SR and RMSD analyses, it can be concluded that the CS algorithm gave a good result in terms of both SR and the number of iterations, especially for the UNIQUAC model. LLE systems containing imidazolium ILs appear to be challenging for PE for higher precision. Further studies should be focused on the performance improvement of the CS algorithm to get result with good precision using low numerical effort. Thus, it can be concluded that the UNIQUAC and the NRTL models, with interaction parameters estimated by the CS algorithm, were able to correlate the LLE data successfully.

7.8 Conclusions

The LLE data for 39 ternary systems including 32 IL-based systems were correlated by the UNIQUAC and the NRTL models. The binary interaction parameters were estimated using the CS algorithm. It has been found that the population size of 20 is sufficient to satisfactorily predict the LLE with high accuracy. %SR analysis suggested that the performance of CS mainly depends on the type of ILs and LLE systems. Especially, ternary systems containing imidazolium ILs

TABLE 7.15

UNIQUAC Binary Interaction Parameters and RMSD Values for Ionic Liquid-Based Ternary Systems

System	Binary Interaction Parameters (K)						RMSD		
No.	A_{12}	A_{21}	A_{13}	A_{31}	A_{23}	A_{32}	(this work)	RMSD (lit.)	Reference
1	−2.19	−152.95	29.59	190.74	−246.23	716.64	0.0058	0.0101	(Santiago et al., 2009)
2	383.19	−297.74	−36.77	460.98	−207.15	421.52	0.0027	0.0038	(Santiago et al., 2009)
3	458.01	−171.18	−129.45	378.98	−155.56	643.72	0.0040	0.0063	(Santiago et al., 2009)
4	521.19	−171.80	−55.47	627.24	−194.46	685.15	0.0019	0.0071	(Santiago et al., 2009)
5	−102.4	274.03	−84.97	470.60	69.84	81.16	0.0054	0.0114	(Santiago et al., 2009)
6	234.72	−29.42	−39.30	367.97	−111.45	173.95	0.0045	0.0064	(Santiago et al., 2009)
7	26.52	155.76	10.41	184.19	−21.85	201.47	0.0049	0.0130	(Santiago et al., 2009)
8	164.99	64.04	−174.63	546.85	14.50	160.35	0.0080	0.0236	(Santiago et al., 2009)
9	224.53	30.55	−157.40	415.66	5.13	159.24	0.0113	0.0246	(Santiago et al., 2009)
10	−119.6	305.63	−58.91	497.37	80.19	−22.81	0.0088	0.0241	(Santiago et al., 2009)
11	473.07	−84.94	680.35	155.92	83.05	135.69	0.0119	0.0276	(Santiago et al., 2009)
12	−223.05	600.06	−18.37	362.53	209.49	−141.47	0.0064	0.0228	(Santiago et al., 2009)
13	124.42	64.50	96.73	137.73	263.34	−93.08	0.0109	0.0300	(Santiago et al., 2009)
14	315.17	−18.63	−39.89	417.74	−53.43	606.50	0.0055	0.0089	(Santiago et al., 2009)
15	231.76	3.71	−9.68	320.92	−67.76	587.51	0.0066	0.0091	(Santiago et al., 2009)
16	160.06	27.14	94.61	142.40	281.60	−73.26	0.0059	0.0203	(Santiago et al., 2009)
17	104.98	72.94	−23.90	327.07	170.11	10.73	0.0051	0.0215	(Santiago et al., 2009)
18	−136.50	392.82	−42.74	206.13	378.99	−190.76	0.0058	0.0243	(Santiago et al., 2009)
19	−210.73	966.00	0.12	120.51	994.64	−278.94	0.0089	0.0406	(Santiago et al., 2009)

(Continued)

TABLE 7.15 (*Continued*)

UNIQUAC Binary Interaction Parameters and RMSD Values for Ionic Liquid-Based Ternary Systems

System	Binary Interaction Parameters (K)						RMSD		
No.	A_{12}	A_{21}	A_{13}	A_{31}	A_{23}	A_{32}	(this work)	RMSD (lit.)	Reference
20	−59.60	201.92	138.04	352.36	211.81	−93.32	0.0041	0.0058	(Revelli et al., 2010)
21	29.38	137.11	131.42	417.88	69.07	131.67	0.0016	0.0042	(Revelli et al., 2010)
22	−438.36	2666.98	2426.07	−104.39	2225.62	−24.24	0.0024*	0.0048	(Rabari & Banerjee, 2014)
23	−315.99	−70.82	2544.94	−312.03	2750.69	−88.10	0.0014*	0.0055	(Rabari & Banerjee, 2014)
24	−179.89	82.36	925.72	−249.73	504.79	−4.65	0.0017	0.0031	(Rabari & Banerjee, 2013)
25	−405.21	775.64	640.95	−189.43	2698.66	−69.92	0.0027*	0.0056	(Rabari & Banerjee, 2013)
26	23.22	157.91	280.22	251.85	−21.95	105.26	0.0011	0.0020	(García et al., 2009)
27	−126.89	613.09	122.98	399.26	58.47	−29.55	0.0016	0.0060	(González et al., 2010)
28	−81.46	429.31	146.08	382.40	−22.88	83.91	0.0010	0.0040	(González et al., 2010)
29	18.00	201.37	86.38	445.91	−10.37	86.24	0.0011	0.0020	(González et al., 2010)
30	668.94	−36.16	599.96	−72.44	95.63	396.95	0.0021	0.0014	(Cháfer et al., 2015)
31	319.61	−105.95	594.27	−86.88	822.94	−219.49	0.0014	0.0017	(Cháfer et al., 2015)
32	237.67	121.16	674.04	−130.94	319.46	78.34	0.0017	0.0015	(Cháfer et al., 2015)
Global RMSD							**0.0056**	**0.0156**	

* lower bound = −3000 and upper bound = +3000.

TABLE 7.16

UNIQUAC Binary Interaction Parameters and RMSD Values for Organic-Solvent-Based Ternary Systems

	Binary Interaction Parameters (K)								
System No.	A_{12}	A_{21}	A_{13}	A_{31}	A_{23}	A_{32}	**RMSD (this work)**	**RMSD (lit.)**	**Reference**
1	−458.86	2871.24	−78.55	1192.58	61.51	75.51	0.0036*	0.0084	(Feng et al., 2015)
2	−184.45	597.67	−16.58	833.18	−3.34	175.78	0.0045	0.0086	(Feng et al., 2015)
3	−328.81	2823.10	−169.54	866.80	−76.79	417.34	0.0062*	0.0070	(Feng et al., 2015)
4	112.27	−47.37	489.39	59.04	−90.84	284.71	0.0015	0.0087	(Luo et al., 2015)
5	522.14	−286.47	420.67	149.94	−42.64	171.08	0.0016	0.0074	(Luo et al., 2015)
6	557.99	−278.58	675.88	53.52	−83.29	299.54	0.0018	0.0042	(Luo et al., 2015)
7	577.05	−301.10	552.05	135.49	−21.14	222.64	0.0011	0.0083	(Luo et al., 2015)
Global RMSD							**0.0036**	**0.0077**	

* lower bound = −3000 and upper bound = +3000.

TABLE 7.17

NRTL Binary Interaction Parameters and RMSD Values for Ionic Liquid-Based Ternary Systems

System No.	Binary Interaction Parameters						RMSD (this work)	RMSD (lit.)	Reference
	τ_{12}	τ_{13}	τ_{21}	τ_{23}	τ_{31}	τ_{32}			
1	28.39	18.57	−3.11	85.37	7.00	2.28	0.0033	0.0067	(Arce et al., 2004b)
2	26.67	19.63	2.62	44.79	9.11	2.97	0.0044	0.0031	(Arce et al., 2004b)
3	−1.00	17.93	3.75	57.90	6.94	2.43	0.0037	0.0058	(Arce et al., 2006b)
4	25.40	20.24	2.47	89.37	7.21	2.99	0.0058	0.0047	(Arce et al., 2006a)
5	97.55	0.81	4.12	2.16	5.04	1.21	0.0122	0.0071	(Alonso et al., 2007)
6	21.34	18.59	4.54	−2.78	6.09	2.68	0.0062	0.0061	(Alonso et al., 2007)
7	27.92	0.32	3.74	3.53	5.57	19.70	0.0121	0.0171	(Alonso et al., 2008)
8	85.64	1.24	3.43	3.37	5.46	80.02	0.0132	0.0231	(Alonso et al., 2008)
9	−0.87	22.20	4.33	25.76	4.86	1.55	0.0174	0.0277	(Alonso et al., 2008)
10	11.72	19.83	21.51	87.99	5.43	8.07	0.0072	0.0180	(Letcher & Reddy, 2005)
11	43.92	23.70	3.63	15.27	10.80	−55.95	0.0128	0.0160	(Letcher & Reddy, 2005)
12	28.32	1.05	21.86	80.23	5.99	0.59	0.0053	0.0150	(Letcher & Reddy, 2005)
13	3.51	36.42	21.38	2.97	4.60	1.48	0.0070	0.0040	(Letcher & Deenadayalu, 2003)
14	−2.61	3.49	7.96	2.74	37.49	35.75	0.0095	0.0050	(Letcher & Reddy, 2004)
15	84.11	72.12	3.34	18.60	8.68	3.32	0.0084	0.0090	(Letcher & Reddy, 2004)
16	2.00	5.44	20.43	1.58	78.70	20.02	0.0085	0.0220	(Letcher & Reddy, 2004)

(*Continued*)

TABLE 7.17 (*Continued*)

NRTL Binary Interaction Parameters and RMSD Values for Ionic Liquid-Based Ternary Systems

	Binary Interaction Parameters								
System No.	τ_{12}	τ_{13}	τ_{21}	τ_{23}	τ_{31}	τ_{32}	RMSD (this work)	RMSD (lit.)	Reference
17	67.19	2.38	2.97	2.87	18.47	–82.02	0.0111	0.0370	(Letcher & Reddy, 2004)
18	20.31	1.41	16.12	22.24	3.79	34.87	0.0041	0.0060	(Letcher & Reddy, 2005)
19	21.13	1.43	16.47	21.80	3.79	33.64	0.0064	0.0040	(Letcher & Reddy, 2005)
20	86.52	4.32	6.07	12.60	35.19	0.69	0.0046	0.0236	(Revelli et al., 2010)
21	0.10	5.34	24.57	1.22	5.63	54.01	0.0036	0.0027	(Revelli et al., 2010)
22	–3.13	74.27	17.31	97.84	8.30	4.32	0.0020	0.0012	(Rabari & Banerjee, 2014)
23	–4.74	32.49	17.64	91.26	6.43	3.80	0.0012	0.0014	(Rabari & Banerjee, 2014)
24	–1.59	21.65	1.16	16.67	5.53	2.91	0.0040	0.0015	(Rabari & Banerjee, 2013)
25	84.25	35.13	–7.19	16.52	8.44	3.16	0.0024	0.0028	(Rabari & Banerjee, 2013)
26	2.24	3.31	43.18	55.11	6.83	1.96	0.0011	0.0018	(García et al., 2009)
27	92.53	1.55	13.83	70.15	5.72	0.32	0.0016	0.0020	(González et al., 2010)
28	86.19	2.12	13.63	79.35	5.29	0.45	0.0010	0.0030	(González et al., 2010)
29	48.84	26.27	13.72	74.38	6.23	0.42	0.0020	0.0020	(González et al., 2010)
30	30.19	0.53	0.21	16.91	9.79	75.05	0.0043	0.0030	(Cháfer et al., 2015)
31	–45.39	22.26	0.27	17.83	9.29	–44.50	0.0024	0.0023	(Cháfer et al., 2015)
32	58.67	21.85	–0.18	4.64	11.41	29.56	0.0026	0.0019	(Cháfer et al., 2015)
Global RMSD							**0.0076**	**0.0139**	

TABLE 7.18

NRTL Binary Interaction Parameters and RMSD Values for Organic-Solvent-Based Ternary Systems

	Binary Interaction Parameters								
System No.	τ_{12}	τ_{21}	τ_{13}	τ_{31}	τ_{23}	τ_{32}	**RMSD (this work)**	**RMSD (lit.)**	**Reference**
1	0.50	18.86	69.19	70.70	5.77	2.20	0.0032	0.0041	(Feng et al., 2015)
2	−75.12	20.88	2.49	1.99	11.28	67.72	0.0041	0.0053	(Feng et al., 2015)
3	−3.21	4.92	6.29	98.67	37.61	2.59	0.0076	0.0162	(Feng et al., 2015)
4	1.56	2.81	65.81	7.71	25.71	2.43	0.0036	0.0081	(Luo et al., 2015)
5	0.51	3.12	13.22	12.35	33.26	3.23	0.0037	0.0054	(Luo et al., 2015)
6	−7.91	2.66	98.17	2.83	20.39	36.13	0.0042	0.0072	(Luo et al., 2015)
7	−2.09	3.23	65.23	82.79	31.15	3.14	0.0058	0.0074	(Luo et al., 2015)
Global RMSD							**0.0048**	**0.0085**	

TABLE 7.19

Comparison of CS with the GA and the PSO Algorithm for Quaternary Systems

				CS		RMSD%		
System	Model	Temp (K)	*i–j*	τ_{ij}	τ_{ji}	CS	GA[a]	PSO[a]
Ethanol (1)	NRTL	293.15	1–2	–49.75	–94.25	0.31	1.267	1.233
Water (2)			1–3	11.21	29.51			
Hexane (3)			1–4	10.98	60.45			
Cyclohexane (4)			2–3	2.62	39.45			
			2–4	1.03	–41.04			
			3–4	73.65	71.04			
Ethanol (1)	NRTL	303.15	1–2	95.13	55.17	0.26		
Water (2)			1–3	30.28	2.52			
Hexane (3)			1–4	–43.75	4.40			
Cyclohexane (4)			2–3	53.28	4.69			
			2–4	8.48	4.55			
			3–4	–1.38	60.03			
Ethanol (1)	NRTL	308.15	1–2	39.69	51.06	0.44		
Water (2)			1–3	42.06	2.47			
Hexane (3)			1–4	37.53	2.39			
Cyclohexane (4)			2–3	39.62	4.25			
			2–4	29.86	6.74			
			3–4	31.87	–44.12			
Ethanol (1)	NRTL	293.15	1–2	71.52	–91.38	0.14	1.032	1.112
Water (2)			1–3	–94.29	–89.42			
Pentane (3)			1–4	17.79	–89.10			
Cyclohexane (4)			2–3	–16.17	4.16			
			2–4	60.84	5.81			
			3–4	–64.59	90.90			

(*Continued*)

TABLE 7.19 (*Continued*)

Comparison of CS with the GA and the PSO Algorithm for Quaternary Systems

System	Model	Temp (K)	*i–j*	CS τ_{ij}	CS τ_{ji}	RMSD% CS	RMSD% GA[a]	RMSD% PSO[a]
Ethanol (1)	NRTL	303.15	1–2	79.21	81.28	0.31		
Water (2)			1–3	36.99	2.37			
Pentane (3)			1–4	70.35	2.50			
Cyclohexane (4)			2–3	32.06	3.63			
			2–4	2.21	50.12			
			3–4	71.95	31.19			
Ethanol (1)	NRTL	308.15	1–2	54.91	38.80	0.33		
Water (2)			1–3	−77.47	2.14			
Pentane (3)			1–4	−92.07	2.66			
Cyclohexane (4)			2–3	88.93	4.11			
			2–4	23.63	4.21			
			3–4	28.67	−29.60			
Ethanol (1)	NRTL	293.15	1–2	86.22	−74.37	0.27	1.021	1.031
Water (2)			1–3	−78.62	−72.31			
Pentane (3)			1–4	52.55	−32.82			
Hexane (4)			2–3	78.20	4.70			
			2–4	−79.86	60.47			
			3–4	−83.83	72.96			

(*Continued*)

TABLE 7.19 (*Continued*)

Comparison of CS with the GA and the PSO Algorithm for Quaternary Systems

				CS		RMSD%		
System	**Model**	**Temp (K)**	***i–j***	τ_{ij}	τ_{ji}	**CS**	**GA**[a]	**PSO**[a]
Ethanol (1)	NRTL	303.15	1–2	−1.85	58.42	0.34		
Water (2)			1–3	3.16	−68.27			
Pentane (3)			1–4	2.98	−7.83			
Hexane (4)			2–3	2.48	2.79			
			2–4	62.65	8.41			
			3–4	−4.39	1.63			
Ethanol (1)	NRTL	308.15	1–2	−94.95	−84.66	0.37		
Water (2)			1–3	11.72	43.44			
Pentane (3)			1–4	13.74	−5.05			
Hexane (4)			2–3	2.12	−71.55			
			2–4	1.29	−90.86			
			3–4	92.89	70.49			

[a] Data from Khansary, M. A & Hallaji Sani, A., *Fluid Phase Equilib.*, 365, 141–145, 2014.

TABLE 7.20

Comparison of CS with the GA and the PSO Algorithm for Quinary Systems

System	Model	Temp (K)	*i–j*	CS τ_{ij}	CS τ_{ji}	RMSD% CS	RMSD% GA[a]	RMSD% PSO[a]
Ethanol (1)	NRTL	303.15	1–2	−47.78	15.78	0.85	2.11	2.07
Water (2)			1–3	−80.29	49.33			
Pentane (3)			1–4	−70.43	28.75			
Hexane (4)			1–5	−56.56	−86.43			
Cyclohexane (5)			2–3	27.71	−42.10			
			2–4	27.35	−43.29			
			2–5	95.84	26.65			
			3–4	−72.19	−81.97			
			3–5	−72.54	−12.76			
			4–5	−69.12	20.29			
Ethanol (1)	NRTL	308.15	1–2	82.87	36.80	0.83		
Water (2)			1–3	64.09	96.85			
Pentane (3)			1–4	−77.30	2.24			
Hexane (4)			1–5	−60.06	25.66			
Cyclohexane (5)			2–3	−7.81	22.34			
			2–4	−91.92	87.86			
			2–5	−52.69	53.04			
			3–4	−92.02	−14.27			
			3–5	−63.84	2.14			
			4–5	3.46	−16.67			

[a] Data from Khansary, M. A & Hallaji Sani, A., *Fluid Phase Equilib.*, 365, 141–145, 2014.

appeared to be most challenging for PE with the NRTL model. RMSD results are extremely encouraging with 0.0053 (UNIQUAC) and 0.0072 (NRTL) deviations in phase compositions which are 63% and 45% better than the global values of 0.0145 and 0.0131 reported in the literature for 371 tie lines. Three ternary systems and one quinary system were selected for the comparison of CS with the GA and the PSO algorithm. Global %RMSD value obtained with CS algorithm was ~0.14%–0.85% compared to ~1.0%–2.0% with the GA and the PSO algorithm. This shows that the CS algorithm is a reliable tool for the estimation of optimum values of binary interaction parameters for IL-based LLE systems and is better in some aspect compared to other optimization techniques.

Nomenclature

$y_i^{(t)}$	*i*th nest current position
α	step size parameter
Rand	random number
S	Step length

Γ	Standard gamma function
p_a	Discovery probability
ζ	Random number
γ_i	Activity coefficient of component *i*
x_i	Mole fraction of component *i*
Φ	Segment fraction in the UNIQUAC model
θ	Area fraction in the UNIQUAC model
z	Lattice coordination number ($z = 10$)
r	Volume parameter in the UNIQUAC parameter
q	Surface parameter in the UNIQUAC parameter
c	Number of components in the LLE system
τ_{ij}	Interaction parameter between components *i* and *j* in the UNIQUAC/NRTL models
A_{ij}	Energy parameter between components *i* and *j* in the UNIQUAC model
α	NRTL non-randomness parameter
m	Number of tie lines
x_{ik}^{l}	Experimental values of mole fraction for component *i* for the *k*th tie line in phase *l*
$\hat{x}_{ik}^{l}$	Predicted values of mole fraction for component *i* for the *k*th tie line in phase *l*
F_{obj}	Objective function
G_{ji}/g_{ji}	Interaction energy parameter

References

Abrams, D. S., & Prausnitz, J. M. (1975). Statistical thermodynamics of liquid mixtures: A new expression for the excess Gibbs energy of partly or completely miscible systems. *AIChE Journal, 21*(1), 116–128. doi:10.1002/aic.690210115.

Alonso, L., Arce, A., Francisco, M., Rodríguez, O., & Soto, A. (2007). Gasoline desulfurization using extraction with [C8mim][BF4] ionic liquid. *AIChE Journal, 53*(12), 3108–3115. doi:10.1002/aic.11337.

Alonso, L., Arce, A., Francisco, M., & Soto, A. (2008). Solvent extraction of thiophene from n-alkanes (C7, C12, and C16) using the ionic liquid [C8mim][BF4]. *The Journal of Chemical Thermodynamics, 40*(6), 966–972. doi:10.1016/j.jct.2008.01.025.

Arce, A., Rodríguez, O., & Soto, A. (2004a). Experimental determination of liquid–liquid equilibrium using ionic liquids: Tert-amyl ethyl ether + ethanol + 1-octyl-3-methylimidazolium chloride system at 298.15 K. *Journal of Chemical & Engineering Data, 49*(3), 514–517. doi:10.1021/je0302147.

Arce, A., Rodríguez, O., & Soto, A. (2004b). Tert-amyl ethyl ether separation from its mixtures with ethanol using the 1-butyl-3-methylimidazolium trifluoromethanesulfonate ionic liquid: Liquid–liquid equilibrium. *Industrial & Engineering Chemistry Research, 43*(26), 8323–8327. doi:10.1021/ie049621h.

Arce, A., Rodríguez, H., & Soto, A. (2006a). Effect of anion fluorination in 1-ethyl-3-methylimidazolium as solvent for the liquid extraction of ethanol from ethyl tert-butyl ether. *Fluid Phase Equilibria, 242*(2), 164–168. doi:10.1016/j.fluid.2006.01.008.

Arce, A., Rodríguez, H., & Soto, A. (2006b). Purification of ethyl tert-butyl ether from its mixtures with ethanol by using an ionic liquid. *Chemical Engineering Journal, 115*(3), 219–223. doi:10.1016/j.cej.2005.10.010.

Aznar, M. (2007). Correlation of (liquid + liquid) equilibrium of systems including ionic liquids. *Brazilian Journal of Chemical Engineering, 24*, 143–149.

Banerjee, T., Singh, M. K., Sahoo, R. K., & Khanna, A. (2005). Volume, surface and UNIQUAC interaction parameters for imidazolium based ionic liquids via polarizable continuum model. *Fluid Phase Equilibria, 234*(1–2), 64–76. doi:10.1016/j.fluid.2005.05.017.

Bhargava, V., Fateen, S. E. K., & Bonilla-Petriciolet, A. (2013). Cuckoo search: A new nature-inspired optimization method for phase equilibrium calculations. *Fluid Phase Equilibria, 337*, 191–200. doi:10.1016/j.fluid.2012.09.018.

Bharti, A., & Banerjee, T. (2015). Enhancement of bio-oil derived chemicals in aqueous phase using ionic liquids: Experimental and COSMO-SAC predictions using a modified hydrogen bonding expression. *Fluid Phase Equilibria, 400*, 27–37. doi:10.1016/j.fluid.2015.04.029.

Bonilla-Petriciolet, A., Fateen, S. E. K., & Rangaiah, G. P. (2013). Assessment of capabilities and limitations of stochastic global optimization methods for modeling mean activity coefficients of ionic liquids. *Fluid Phase Equilibria, 340*, 15–26. doi:10.1016/j.fluid.2012.12.002.

Bonilla-Petriciolet, A., Rangaiah, G. P., & Segovia-Hernández, J. G. (2010). Evaluation of stochastic global optimization methods for modeling vapor–liquid equilibrium data. *Fluid Phase Equilibria*, 287(2), 111–125. doi:10.1016/j.fluid.2009.09.021.

Bonilla-Petriciolet, A., & Segovia-Hernández, J. G. (2010). A comparative study of particle swarm optimization and its variants for phase stability and equilibrium calculations in multicomponent reactive and non-reactive systems. *Fluid Phase Equilibria, 289*(2), 110–121. doi:10.1016/j.fluid.2009.11.008.

Cháfer, A., de la Torre, J., Font, A., & Lladosa, E. (2015). Liquid–liquid equilibria of water + ethanol + 1-butyl-3-methylimidazolium bis(trifluoromethanesulfonyl)imide ternary system: Measurements and correlation at different temperatures. *Journal of Chemical & Engineering Data, 60*(8), 2426–2433. doi:10.1021/acs.jced.5b00301.

Fateen, S.-E. K., & Bonilla-Petriciolet, A. (2014a). On the effectiveness of the nature-inspired metaheuristic algorithms for performing phase equilibrium thermodynamic calculations. *The Scientific World Journal, 2014*, 2014374510.1. doi:10.1155/2014/374510.

Fateen, S.-E. K., & Bonilla-Petriciolet, A. (2014b). A note on effective phase stability calculations using a gradient-based cuckoo search algorithm. *Fluid Phase Equilibria, 375*, 360–366. doi:10.1016/j.fluid.2014.05.009.

Fateen, S.-E. K., Bonilla-Petriciolet, A., & Rangaiah, G. P. (2012). Evaluation of covariance matrix adaptation evolution strategy, shuffled complex evolution and firefly algorithms for phase stability, phase equilibrium and chemical equilibrium problems. *Chemical Engineering Research and Design, 90*(12), 2051–2071. doi:10.1016/j.cherd.2012.04.011.

Feng, Y., Yang, E., Dang, L., & Wei, H. (2015). Liquid–liquid phase equilibrium for ternary mixtures of formamide (or ethylene glycol, or monoethanolamine) + indole + 2-methylnaphthalene at 308.15 K. *Fluid Phase Equilibria, 398,* 10–14. doi:10.1016/j.fluid.2015.04.001.

Fernández-Vargas, J. A., Bonilla-Petriciolet, A., & Segovia-Hernández, J. G. (2013). An improved ant colony optimization method and its application for the thermodynamic modeling of phase equilibrium. *Fluid Phase Equilibria, 353,* 121–131. doi:10.1016/j.fluid.2013.06.002.

Ferrari, J. C., Nagatani, G., Corazza, F. C., Oliveira, J. V., & Corazza, M. L. (2009). Application of stochastic algorithms for parameter estimation in the liquid–liquid phase equilibrium modeling. *Fluid Phase Equilibria, 280*(1–2), 110–119. doi:10.1016/j.fluid.2009.03.015.

Fister, I., Jr. Fister, D., & Fister, I.A. (2013). Comprehensive review of cuckoo search, variants and Hybrids. *Int. J. Mathematical Modelling and Numerical Optimisation, 4,* 387– 409. doi: 10.1504/IJMMNO.2013.059205.

García, J., Fernández, A., Torrecilla, J. S., Oliet, M., & Rodríguez, F. (2009). Liquid–liquid equilibria for {hexane + benzene + 1-ethyl-3-methylimidazolium ethylsulfate} at (298.2, 313.2 and 328.2) K. *Fluid Phase Equilibria, 282*(2), 117–120. doi:10.1016/j.fluid.2009.05.006.

González, E. J., Calvar, N., González, B., & Domínguez, Á. (2010). Liquid extraction of benzene from its mixtures using 1-ethyl-3-methylimidazolium ethylsulfate as a solvent. *Journal of Chemical & Engineering Data, 55*(11), 4931–4936. doi:10.1021/je100508y.

Huang, Ch., Chung, P., Tseng, I., & Lee, L. (2010). Measurements and correlations of liquid-liquid-equilibria of the mixtures consisting of ethanol, water, pentane, hexane, and cyclohexane. *The Open Thermodynamics Journal, 4,* 102–118.

Jaime-Leal, J. E., Bonilla-Petriciolet, A., Bhargava, V., & Fateen, S. E. K. (2015). Nonlinear parameter estimation of e-NRTL model for quaternary ammonium ionic liquids using Cuckoo Search. *Chemical Engineering Research and Design, 93,* 464–472. doi:10.1016/j.cherd.2014.06.014.

Kabouche, A., Boultif, A., Abidi, A., & Gherraf, N. (2012). Interaction parameter estimation in liquid–liquid phase equilibrium modeling using stochastic and hybrid algorithms. *Fluid Phase Equilibria, 336,* 113–121. doi:10.1016/j.fluid.2012.09.002.

Karaboga, D., & Basturk, B. (2007). A powerful and efficient algorithm for numerical function optimization: artificial bee colony (ABC) algorithm. *Journal of Global Optimization, 39*(3), 459–471. doi:10.1007/s10898-007-9149-x.

Khansary, M. A, & Sani, A. H. (2014). Using genetic algorithm (GA) and particle swarm optimization (PSO) methods for determination of interaction parameters in multicomponent systems of liquid–liquid equilibria. *Fluid Phase Equilibria, 365,* 141–145. doi:10.1016/j.fluid.2014.01.016.

Letcher, T. M., & Deenadayalu, N. (2003). Ternary liquid–liquid equilibria for mixtures of 1-methyl-3-octyl-imidazolium chloride + benzene + an alkane at T = 298.2 K and 1 atm. *The Journal of Chemical Thermodynamics, 35*(1), 67–76. doi:10.1016/S0021-9614(02)00300-2.

Letcher, T. M., & Reddy, P. (2004). Ternary liquid–liquid equilibria for mixtures of 1-hexyl-3-methylimidozolium (tetrafluoroborate or hexafluorophosphate) + ethanol + an alkene at T = 298.2 K. *Fluid Phase Equilibria, 219*(2), 107–112. doi:10.1016/j.fluid.2003.10.012.

Letcher, T. M., & Reddy, P. (2005). Ternary (liquid + liquid) equilibria for mixtures of 1-hexyl-3-methylimidazolium (tetrafluoroborate or hexafluorophosphate) + benzene + an alkane at T = 298.2 K and p = 0.1 MPa. *The Journal of Chemical Thermodynamics, 37*(5), 415–421. doi:10.1016/j.jct.2004.05.001.

Luo, L., Liu, D., Li, L., & Chen, Y. (2015). Phase equilibria of (water + propionic acid or butyric acid + 2-methoxy-2-methylpropane) ternary systems at 298.2 K and 323.2 K. *Fluid Phase Equilibria, 403,* 30–35. doi:10.1016/j.fluid.2015.06.003.

Merzougui, A., Hasseine, A., Kabouche, A., & Korichi, M. (2011). LLE for the extraction of alcohol from aqueous solutions with diethyl ether and dichloromethane at 293.15 K, parameter estimation using a hybrid genetic based approach. *Fluid Phase Equilibria, 309*(2), 161–167. doi:10.1016/j.fluid.2011.07.011.

Merzougui, A., Hasseine, A., & Laiadi, D. (2012). Application of the harmony search algorithm to calculate the interaction parameters in liquid–liquid phase equilibrium modeling. *Fluid Phase Equilibria, 324,* 94–101. doi:10.1016/j.fluid.2012.03.029.

Nanda, S. J., & Panda, G. (2014). A survey on nature inspired metaheuristic algorithms for partitional clustering. *Swarm and Evolutionary Computation, 16,* 1–18. doi:10.1016/j.swevo.2013.11.003.

Rabari, D., & Banerjee, T. (2013). Biobutanol and n-propanol recovery using a low density phosphonium based ionic liquid at T = 298.15 K and p = 1 atm. *Fluid Phase Equilibria, 355,* 26–33. doi:10.1016/j.fluid.2013.06.047.

Rabari, D., & Banerjee, T. (2014). Experimental and theoretical studies on the effectiveness of phosphonium-based ionic liquids for butanol removal at T = 298.15 K and p = 1 atm. *Industrial & Engineering Chemistry Research, 53*(49), 18935–18942. doi:10.1021/ie500833h.

Renon, H., & Prausnitz, J. M. (1968). Local compositions in thermodynamic excess functions for liquid mixtures. *AIChE Journal, 14*(1), 135–144. doi:10.1002/aic.690140124.

Revelli, A.-L., Mutelet, F., & Jaubert, J.-N. (2010). Extraction of benzene or thiophene from n-heptane using ionic liquids. NMR and thermodynamic study. *The Journal of Physical Chemistry B, 114*(13), 4600–4608. doi:10.1021/jp911978a.

Sahoo, R. K., Banerjee, T., Ahmad, S. A., & Khanna, A. (2006). Improved binary parameters using GA for multi-component aromatic extraction: NRTL model without and with closure equations. *Fluid Phase Equilibria, 239*(1), 107–119. doi:10.1016/j.fluid.2005.11.006.

Santiago, R. S., Santos, G. R., & Aznar, M. (2009). UNIQUAC correlation of liquid–liquid equilibrium in systems involving ionic liquids: The DFT–PCM approach. *Fluid Phase Equilibria, 278*(1–2), 54–61. doi:10.1016/j.fluid.2009.01.002.

Seader, J. D., & Henley, E. J. (2006). Separation Process Principles; John Wiley & Sons: New York.

Singh, M. K., Banerjee, T., & Khanna, A. (2005). Genetic algorithm to estimate interaction parameters of multicomponent systems for liquid–liquid equilibria. *Computers & Chemical Engineering, 29*(8), 1712–1719. doi:10.1016/j.compchemeng.2005.02.020.

Srinivas, M., & Rangaiah, G. P. (2007). Differential evolution with tabu list for global optimization and its application to phase equilibrium and parameter estimation problems. *Industrial & Engineering Chemistry Research, 46*(10), 3410–3421. doi:10.1021/ie0612459.

Varma, N. R., Ramalingam, A., & Banerjee, T. (2011). Experiments, correlations and COSMO-RS predictions for the extraction of benzothiophene from n-hexane using imidazolium-based ionic liquids. *Chemical Engineering Journal, 166*(1), 30–39. doi:10.1016/j.cej.2010.09.015.

Vatani, M., Asghari, M., & Vakili-Nezhaad, G. (2012). Application of genetic algorithm to the calculation of parameters for NRTL and two-suffix Margules models in ternary extraction ionic liquid systems. *Journal of Industrial and Engineering Chemistry, 18*(5), 1715–1720. doi:10.1016/j.jiec.2012.03.008.

Yang, X. S. (2010). Engineering optimization: An introduction with metaheuristic applications; John Wiley & Sons: Hoboken, NJ.

Yang, X.-S. (2014). Nature-Inspired Optimization Algorithms; Elsevier: Waltham, MA.

Yang, X.-S., & Deb, S. (2009). Cuckoo search via Levy flights. In Proceedings of World Congress on Nature & Biologically Inspired Computing (NaBIC 2009), December 9–11, 2009, Coimbatore, India; IEEE Publications: New York, 2009; pp. 210– 214.

Yang, X.-S., & Deb, S. (2013). Multiobjective cuckoo search for design optimization. *Computers & Operations Research, 40*(6), 1616–1624. doi:10.1016/j.cor.2011.09.026.

Yang, X.-S., & Deb, S. (2014). Cuckoo search: Recent advances and applications. *Neural Computing and Applications, 24*(1), 169–174. doi:10.1007/s00521-013-1367-1.

Zhang, H., Fernández-Vargas, J. A., Rangaiah, G. P., Bonilla-Petriciolet, A., & Segovia-Hernández, J. G. (2011). Evaluation of integrated differential evolution and unified bare-bones particle swarm optimization for phase equilibrium and stability problems. *Fluid Phase Equilibria, 310*(1–2), 129–141. doi:10.1016/j.fluid.2011.08.002.

Zhang, H., Rangaiah, G. P., & Bonilla-Petriciolet, A. (2011). Integrated differential evolution for global optimization and its performance for modeling vapor–liquid equilibrium data. *Industrial & Engineering Chemistry Research, 50*(17), 10047–10061. doi:10.1021/ie200819p.

Index

Note: Page numbers followed by f and t refer to figures and tables, respectively.

N

O

P

Q

R